Anesthesia for Lung Transplant

Marcus Salvatori • Alexander Huang
Editors

Anesthesia for Lung Transplant

Springer

Editors
Marcus Salvatori
Department of Anesthesiology and Pain Medicine / Institute of Medical Science, Temerty Faculty of Medicine, University of Toronto
Toronto, ON, Canada

Alexander Huang
Department of Anesthesiology and Pain Medicine / Institute of Medical Science, Temerty Faculty of Medicine, University of Toronto
Toronto, ON, Canada

ISBN 978-3-032-06111-9 ISBN 978-3-032-06112-6 (eBook)
https://doi.org/10.1007/978-3-032-06112-6

This Springer imprint is published by the registered company Springer Nature Switzerland AG
The registered company address is: Gewerbestrasse 11, 6330 Cham, Switzerland

To the patients and their families who challenge us to be better clinicians and compassionate care givers

Preface

Since the first successful lung transplant by Dr. Joel Cooper in 1983, the Toronto Lung Transplant Program (TLTP) has been at the forefront of ingenuity and advancement in lung transplantation. The last decade has seen explosive growth, driven in large part by enhanced preservation techniques, ex vivo lung perfusion (EVLP), intraoperative extracorporeal membrane oxygenation (ECMO) support, and advancements in post-operative care. As a result of these successes, we are now able to help older patients with more advanced disease. Although the COVID pandemic slowed the rates of lung transplantation worldwide, it has also generated an emerging class of patients with post-COVID lung disease, some of whom benefit from transplant. The field of anesthesia for lung transplantation continues to evolve to meet these new and dynamic challenges.

This book reflects the current practice of the Thoracic Anesthesia Group at the Toronto General Hospital (TGH), part of the University Health Network (UHN). We sought to create a practical, evidence-based guide for fellows and anesthesiologists providing anesthesia for lung transplantation. The layout follows a "*Frequently Asked Questions*" (FAQ) format in order to address the nuances characteristic of the diverse patient population. Wherever possible, answers are supported by research from the lung transplantation and thoracic anesthesia literature.

The information provided in this book should be used together with sound clinical judgment to meet the specific needs of individual patients. Institutional practices and clinician preferences may vary. We sincerely hope this handbook helps to build a foundation for better clinical knowledge and practice.

Marcus Salvatori
Alexander Huang

Contributing Authors

Dr. Mika Hamilton	Critical Care Physician
Dr. Sarah McDonald	Anesthesiologist
Dr. Luca Orsini	Critical Care Research Fellow
Dr. Dan Sellers	Anesthesiologist
Dr. Serena Shum	Anesthesiologist

Reviewers

Dr. Lorenzo del Sorbo	Critical Care Physician
Dr. Laura Donahoe	Surgeon
Dr. Rita Katznelson	Anesthesiologist
Dr. Shaf Keshavjee	Surgeon
Dr. Martin Ma	Anesthesiologist
Dr. Karen McRae	Anesthesiologist
Dr. Leonid Minkovich	Anesthesiologist
Dr. Peter Slinger	Anesthesiologist
Dr. Marcin Wasowicz	Anesthesiologist
Dr. Chris Yarnell	Critical Care Physician
Dr. Kazuhiro Yasufuku	Surgeon
Dr. Jonathan Yeung	Surgeon

This page lists current contributors to "Anesthesia for Lung Transplantation" edited by Dr. Marcus Salvatori and Dr. Alexander Huang. The editors wish to thank all contributors for their work

Disclaimer

The content of the book represents the work of the authors and the contributors and not necessarily the University Health Network. Although the authors have made every effort to ensure that the information in this book was correct at press time, neither the authors nor contributors will be liable for damages arising out of or in connection with the use of this book.

Contents

About the Editors

Marcus Salvatori is a staff anesthesiologist and Director of Perioperative Echocardiography at the University Health Network-Toronto General Hospital. He obtained a Master of Science from the University of Toronto Department of Physiology (2005) and a Bachelor of Medicine and Surgery from Newcastle University School of Medicine, Newcastle upon Tyne, UK (2013). He completed his anesthesia residency at the University of Toronto and obtained his Fellowship, Royal College of Physicians of Canada (FRCPC) in Anesthesiology in 2019. He completed a cardiovascular anesthesia and critical care medicine fellowship at the Toronto General Hospital, University Health Network. His areas of interest include adult congenital heart disease, aortic surgery, solid organ transplantation, and transesophageal echocardiography.

Alexander Huang is a staff anesthesiologist and Director of Thoracic Anesthesia at the University Health Network-Toronto General Hospital. He received his MD degree from the University of Toronto (2011) and completed his anesthesia residency at the University of Toronto (2016). He completed fellowship training in cardiac and thoracic anesthesia at University Health Network-Toronto General Hospital. His clinical areas of interest include thoracic anesthesia, lung transplantation, perioperative blood conservation, and acute and transitional pain medicine.

Abbreviations

A1AT	Alpha-1 antitrypsin
ABG	Arterial blood gas
AMR	Antibody-mediated rejection
ARDS	Acute respiratory distress syndrome
ATG	Antithymocyte antiglobulin
CCO	Continuous cardiac output
CF	Cystic fibrosis
CLABSI	Central line associated bloodstream infections
CLAD	Chronic lung allograft dysfunction
CMV	Cytomegalovirus
CO	Cardiac output
COVID	Coronavirus
CPP	Cerebral perfusion pressure
CTEPH	Chronic thromboembolic pulmonary hypertension
CVC	Central venous catheter
CVP	Central venous pressure
DAA	Direct acting antiviral
DBD	Donation after brain death (see NDD)
DCD	Donation after circulatory death
DLETT	Double-lumen endotracheal tube
DLT	Double lung transplant
DSA	Donor-HLA specific antibodies
ECD	Extended criteria donors
ECMO	Extracorporeal membrane oxygenation
ECLS	Extracorporeal life support
EDV	End diastolic volume
$EtCO_2$	End tidal carbon dioxide
ESPB	Erector spinae plane block
EVLP	Ex vivo lung perfusion
FFP	Fresh frozen plasma
FIO_2	Fraction of inspired oxygen
HCO_3	Bicarbonate
HCV	Hepatitis C Virus
HPV	Hypoxic pulmonary vasoconstriction

ICO	Intermittent cardiac output
IJ	Internal jugular
iNO	Inhaled nitric oxide
ISHLT	International Society of Heart and Lung Transplantation
IVIG	Intravenous immune globulin
IV-PCA	Intravenous patient-controlled analgesia
LAS	Lung allocation score
MAID	Medical assistance in dying
MAC	Minimum alveolar concentration
Mg	Magnesium
MPA	Mycophenolic acid
NIRS	Near infrared spectroscopy
NDD	Neurological determination of death (see DBD)
NPOD	Non-perfused organ donors
OLV	One lung ventilation
QoL	Quality of life
PaO_2	Partial pressure of oxygen in arterial blood
PA	Pulmonary artery
PAC	Pulmonary artery catheter
PAP	Pulmonary artery pressure
PCV	Pressure control ventilation
PGD	Primary graft dysfunction
PLEX	Plasma exchange
ppFEV1	Percent predicted FEV1
PVB	Paravertebral block
PVR	Pulmonary vascular resistance
PHTN	Pulmonary hypertension
RV	Right ventricle
RVEF	Right ventricular ejection fraction
SAPB	Serratus anterior plane block
SLT	Single lung transplant
SLETT	Single lumen endotracheal tube
SvO_2	Mixed venous oxygen saturation
TGH	Toronto General Hospital
TGLN	Trillium Gift of Life Network
TIVA	Total intravenous anesthesia
TLTP	Toronto Lung Transplant Program
TPS	Transitional pain service
uDCD	Uncontrolled donation after circulatory death
UNOS	United network on organ sharing
U/S	Ultrasound
VAP	Ventilator associated pneumonia
VCV	Volume control ventilation
WIT	Warm ischemic time

Donors and Procurement

1

Abstract

Successful lung transplantation depends on careful donor selection and meticulous procurement, preservation, and transport. While donation after brain death (DBD) remains the standard, donation after circulatory death (DCD) has emerged as a viable alternative, with comparable results in experienced centers. Procurement protocols involving lung-protective ventilation, hormonal resuscitation, and judicious fluid management are critical to preserving graft function. Advances in organ preservation, particularly static cold storage at 10 °C, have extended safe ischemic times without compromising outcomes. Donor-specific factors, including age, smoking history, diabetes, and infection risk, continue to influence both graft viability and recipient survival.

1.1 Does Donor Type Affect Outcome?

There are two main donor types for lung transplants: (1) donation after brain death (DBD) also known as neurological determination of death (NDD) and (2) donation after cardiac (or circulatory) death (DCD). DBD donors are the more common type for lung transplant grafts worldwide, as they allow for a planned procurement with minimal ischemic time [1] (Fig. 1.1).

DCD lung transplants have become more common since the mid-2000s, although the first DCD lung transplant was performed at the University of Wisconsin in 1995 [2]. Canada adopted DCD donor transplantation in 2006 [3]. Compared to DBD, DCD involves additional considerations, including ethical concerns that vary by culture, as well as prolonged warm ischemic time. Warm ischemic time (WIT) begins at withdrawal of life-sustaining therapies and ends with administration of cold perfusate during the retrieval surgery. The International Society of Heart and Lung Transplantation (ISHLT) registry defines several time periods involved in the

M. Salvatori, A. Huang (eds.), *Anesthesia for Lung Transplant*,
https://doi.org/10.1007/978-3-032-06112-6_1

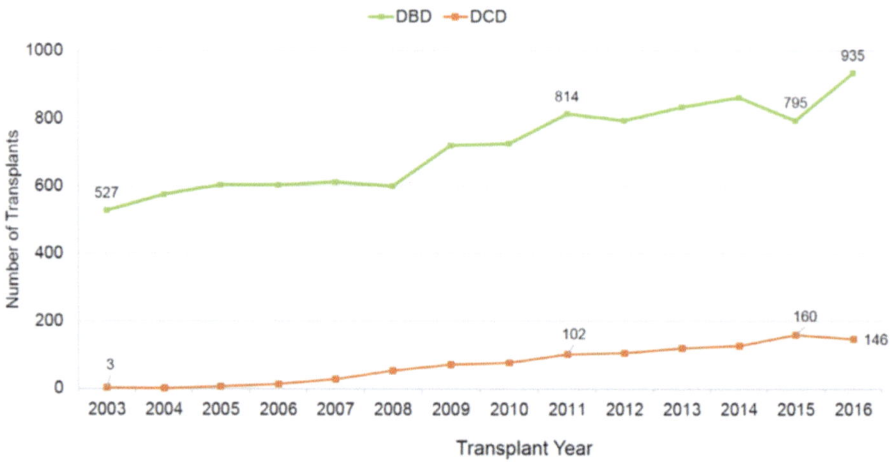

Fig. 1.1 Annual number of DBD and DCD lung transplants as per the ISHLT Registry, 2003–2016 [1]. (DBD, donation after brain death; DCD, donation after circulatory death; ISHLT, International Society of Heart and Lung Transplantation. *J Heart Lung Transplant* 2019;38:1235–1245)

Table 1.1 DCD time periods [4]

T0	Withdrawal of life-sustaining therapies OR euthanasia
T1	Oxygen saturation < 80%
T2	Systolic blood pressure < 50 mmHg
T3	Cessation of cardiac output/asystole
T4	Resumed lung inflation/ventilation
T5	Start of pulmonary flush

DCD donation process [4] (Table 1.1). The concept of "functional warm ischemic time" encompasses the period where meaningful organ perfusion ends, which begins at T2.

In 1995, Kootstra and colleagues [5] published the first categorization of DCD donors, the Maastricht categories, to delineate the ischemic etiologies and circumstances surrounding DCD donation (Table 1.2). Typically, DCD donation is limited to Category III and IV donors (III > IV). The categories have been modified several times, most recently in 2013 [6]. Category V was introduced to reflect organ donation after situations such as medical assistance in dying (MAID).

Initial concerns regarding DCD donation included theories that WIT would lead to worse primary graft dysfunction (PGD), increased rates of rejection, and increased risk for mortality. However, a retrospective study utilizing data from ISHLT registry revealed that there are no differences in survival between DBD and DCD recipients [1]. Several small studies show that DBD and DCD transplants have comparable rates of PGD and acute rejection [8–10].

Uncontrolled donation after circulatory death (uDCD, Maastricht categories I and II) is an emerging area of interest, although the literature is currently limited to small series and case reports. The largest series published thus far examined 38 uDCD cases [11]. When compared to DBD donors, several outcomes including ICU length

Table 1.2 Modified Maastricht Classification [7]

Category	Circumstances	Description
Category I Uncontrolled	Found dead IA. Out of hospital IB. in hospital	Sudden, unexpected cardiac arrest *without* any attempt at resuscitation
Category II Uncontrolled	Witnessed cardiac arrest IIA. Out of hospital IIB. In hospital	Sudden, unexpected cardiac arrest *with an unsuccessful* attempt at resuscitation
Category III Controlled	Withdrawal of life-sustaining therapy	Planned withdrawal of life-sustaining therapy
Category IV Controlled/ uncontrolled	Cardiac arrest in a brain-dead patient	Sudden cardiac arrest after brain death diagnosis
Category V Uncontrolled	Cardiac arrest in an in-hospital patient	*No resuscitation effort*, includes medically assisted death

of stay, hospital length of stay, PGD grade III, and chronic graft dysfunction were equal. However, overall survival was significantly lower in the uDCD transplants. NPOD donors represent a variation of uDCD, where after declaration of death following uncontrolled cardiac arrest, no reinstitution of circulation (i.e., chest compressions) is performed. Lung recovery in NPODs must occur within 3 hours of death, and all NPOD grafts are committed to ex-vivo lung perfusion (EVLP). The early North American/Toronto experience has been recently published, reporting 5 cases of uDCD/NPOD lung transplantation, with positive results [12].

1.2 What Is the Donor Profile at TGH?

Between 2017 and 2021, 69% of lung transplants involved DBD, while 31% were DCD. Of the transplanted DCD donors, 2% were NPOD, and 16% originated from MAID donors.

1.3 What Lung Protective Measures Are Taken During Procurement?

In DBD donors, the use of lung protective ventilation (6-8 cc/kg tidal volume, PEEP 8-10cmH2O, apnea testing with CPAP, closed circuit suctioning) significantly improves the proportion of lungs accepted for transplantation [13], but does not influence post-transplant outcomes [14, 15].

Endocrine support with vasopressin and corticosteroids ± thyroid hormone is standard practice for DBD donors and is associated with improved rates of graft recovery [16]. Administration of high-dose corticosteroids (15 mg/kg IV) following declaration of brain death and every 24 hrs thereafter improves PF (PaO2:FiO_2) ratios in donor lungs [17].

Fluid management is debated for all stages of the lung transplantation process (donor preparation, transplantation, post-transplant care). Judicious, goal-directed fluid therapy balances organ perfusion versus the accumulation of extravascular lung water. Maintaining a donor CVP < 10 mmHg improves rates of graft recovery [16], using diuresis if necessary [18].

1.4 What Is the Maximum Ischemic Time for Donor Grafts?

After lungs are procured, they are cold-flushed with a lung-preservation solution, inflated with oxygen, and stored for transport [19]. Conventionally, grafts were stored in an ice cooler at 4 °C. Although the maximum "ice storage" preservation time is unknown, ischemic times over 8 h are thought to be a risk factor for primary graft dysfunction (PGD) (see Chap. 13).

Challenging convention, recent studies from the TLTP have demonstrated the safe prolongation of ischemic time using cold static preservation at 10°C [19]. Grafts stored at 10 °C show improved inflammatory profile, mitochondrial health, and cellular membrane function [19]. A prospective, multicenter trial from the TLTP showed that lungs could be preserved at 10 °C for 12–14 hours with no meaningful differences in PGD, postoperative extracorporeal membrane oxygenation (ECMO), intensive care stay, hospital stay, or 1-year survival [19].

1.5 Which Donor Characteristics Affect Outcome?

This is a complex topic which comprises donor factors, recipient factors, and factors resulting from the interaction between donor and recipient. For a full discussion, please see the relevant publications by Diamond et al. [20] and Di Nardo et al. [21].

The ideal donor criteria, based on expert consensus, are listed in [22]. Donors that deviate from these criteria are termed extended criteria donors (ECDs). There is no consensus regarding the eligibility of ECD lungs, and factors such as donor age, comorbidities, and infection influence the decision.

Ideal donor criteria [22]

Age < 55
ABO compatible
Clear chest radiograph
PaO_2 > 300 mmHg on 100% FiO_2 and PEEP 5 cm H_2O
Smoking history <20 pack years
Absence of significant chest trauma
No evidence of aspiration/sepsis
No prior cardiopulmonary surgery
Gram stain of sputum/BAL negative
Acceptable bronchoscopy without evidence of aspiration or purulence
No history of significant lung disease

Size matching

Age The evidence supporting the ideal donor cut-off of 55 years is mixed. In a retrospective study of the United Network on Organ Sharing (UNOS) thoracic database, lungs from donors >60 were associated with lower 5-year survival (44% vs. 52%, $p < 0.001$) [23]. However, a separate retrospective analysis found no difference in 1-year graft failure rates among donors aged 55–64 compared to those aged 18–54 [24]. Additionally, reports have shown that the use of donors aged 55–64 years of age was not a risk factor for 1- and 3-year mortality, and the TLTP has shown that advanced donor age (age $\geq$ 70) was not associated with worse overall and CLAD-free survival [25].

Smoking Ideal donors may have a smoking history of less than 20 pack years, although patients with smoking-related lung disease (i.e., COPD) are excluded. In a retrospective review of 1295 DBD lung transplants, recipients of lungs from donors with positive smoking histories had worse 3-year survival after transplantation [26]. The effect was exposure-dependent, with worse survival for lungs from donors who smoked >20 cigarettes per day. However, receiving lungs from a donor with a positive smoking history still conferred better outcomes than remaining on the waiting list.

Diabetes Endocrine diseases impact respiratory physiology and function [27]. Diabetes alters lung volumes, diffusion capacity, central control of ventilation, bronchomotor tone, and neuroadrenergic bronchial innervation [28]. Lungs from diabetic donors are associated with higher overall mortality [29] and worse 1-, 3-, and 5-year survival [30, 31]. The risk appears highest for single lung transplants from diabetic donors [32]. The mechanism by which donor diabetes influences survival is unclear, but hyperglycemic microangiopathic changes may impair diffusion capacity [28]. The pathophysiology is likely exacerbated by post-transplant diabetes, which develops in up to 10% of recipients [33].

Infection With the emergence of direct-acting antivirals (DAAs), grafts from hepatitis C virus (HCV)-infected donors are now eligible for transplantation. Ultraviolet irradiation during EVLP reduces HCV viral loads, and recipients are treated with DAAs post-transplant. HCV-donor grafts show comparable early and intermediate clinical outcomes to non-HCV lungs. However, lungs from cytomegalovirus (CMV)-positive donors have higher overall mortality [29] and post-transplant CMV infection increases risk of death, acute rejection, and development of bronchiolitis obliterans [34]. Nucleic acid testing (NAT) is routinely performed to screen for HBV, HCV, and HIV.

All potential donors are routinely screened for bacterial infection. Donor infection is common, with bacterial growth in 50–63% of potential donors [35, 36]. Fortunately, donor-to-host transmission is infrequent (<10%) due to routine broad-spectrum antibiotic prophylaxis and rapid culture and sensitivity assays [35, 36].

Positive donor cultures do not impact 30-day mortality or PGD-3 at 24 and 72 hrs, but are associated with prolonged mechanical ventilation [35].

References

1. Van Raemdonck D, et al. Donation after circulatory death in lung transplantation-five-year follow-up from ISHLT registry. J Heart Lung Transplant. 2019;38(12):1235–45.
2. Love RB SJ, Chomiak PN, Pellett JR, Mentzer RM. Successful lung transplantation using a non-heart-beating donor (abstract). J Heart Lung Transplant 1995;14.
3. Hernadez-Alejandro R, et al. Organ donation after cardiac death: donor and recipient outcomes after the first three years of the Ontario experience. Can J Anaesth. 2011;58(7):599–605.
4. Cypel M, et al. International Society for Heart and Lung Transplantation donation after circulatory death registry report. J Heart Lung Transplant. 2015;34(10):1278–82.
5. Kootstra G, et al. Categories of non-heart-beating donors. Transplant Proc. 1995;27(5):2893–4.
6. Thuong M, et al. New classification of donation after circulatory death donors definitions and terminology. Transpl Int. 2016;29(7):749–59.
7. Pionnier Y, Darius T, Penaloza A, et al. Solid organ transplantation originating from uncontrolled donation after circulatory death in Europe: a narrative review. Scand J trauma Resusc. Emerg Med. 2024;32:130. This article is licensed under CC BY 4.0
8. Inci I, et al. Lung transplantation with controlled donation after circulatory death donors. Ann Thorac Cardiovasc Surg. 2018;24(6):296–302.
9. Krutsinger D, et al. Lung transplantation from donation after cardiocirculatory death: a systematic review and meta-analysis. J Heart Lung Transplant. 2015;34(5):675–84.
10. Barbero C, et al. Lung donation after circulatory determined death: a single-Centre experience. Eur J Cardiothorac Surg. 2019;55(2):309–15.
11. Valdivia D, et al. Expanding the horizons: uncontrolled donors after circulatory death for lung transplantation-first comparison with brain death donors. Clin Transpl. 2019;33(6):e13561.
12. Healey A, et al. Initial lung transplantation experience with uncontrolled donation after cardiac death in North America. Am J Transplant. 2020;20(6):1574–81.
13. Ball IM, et al. Management of the neurologically deceased organ donor: a Canadian clinical practice guideline. CMAJ. 2020;192(14):E361–E69.
14. Anderson TA, et al. Anesthetic considerations in organ procurement surgery: a narrative review. Can J Anaesth. 2015;62(5):529–39.
15. Mascia L, et al. Effect of a lung protective strategy for organ donors on eligibility and availability of lungs for transplantation: a randomized controlled trial. JAMA. 2010;304(23):2620–7.
16. Abdelnour T, Rieke S. Relationship of hormonal resuscitation therapy and central venous pressure on increasing organs for transplant. J Heart Lung Transplant. 2009;28(5):480–5.
17. Follette DM, et al. Improved oxygenation and increased lung donor recovery with high-dose steroid administration after brain death. J Heart Lung Transplant. 1998;17(4):423–9.
18. Wood KE, et al. Care of the potential organ donor. N Engl J Med. 2004;351(26):2730–9.
19. Cypel M, et al. Lungs preserved on ice or in a refrigerator? Prolonged static lung storage at 10 degrees C. Ann Thorac Surg. 2023;115(5):1095–7.
20. Diamond JM, et al. Clinical risk factors for primary graft dysfunction after lung transplantation. Am J Respir Crit Care Med. 2013;187(5):527–34.
21. Di Nardo M, et al. Postoperative Management of Lung Transplant Recipients in the intensive care unit. Anesthesiology. 2022;136(3):482–99.
22. Orens JB, et al. A review of lung transplant donor acceptability criteria. J Heart Lung Transplant. 2003;22(11):1183–200.
23. Whited WM, et al. Does donor age and double versus single lung transplant affect survival of young recipients? Ann Thorac Surg. 2018;105(1):235–41.

24. Baldwin MR, et al. Donor age and early graft failure after lung transplantation: a cohort study. Am J Transplant. 2013;13(10):2685–95.
25. Saddoughi SA, et al. Extending the age criteria of lung transplant donors to 70+ years old does not significantly affect recipient survival. J Thorac Cardiovasc Surg. 2023;167:861.
26. Bonser RS, et al. Effect of donor smoking on survival after lung transplantation: a cohort study of a prospective registry. Lancet. 2012;380(9843):747–55.
27. Lencu C, et al. Respiratory manifestations in endocrine diseases. Clujul Med. 2016;89(4):459–63.
28. Pitocco D, et al. The diabetic lung--a new target organ? Rev Diabet Stud. 2012;9(1):23–35.
29. Reyes KG, et al. Guidelines for donor lung selection: time for revision? Ann Thorac Surg 2010;89(6):1756–64; discussion 64–5.
30. Christie JD, et al. The registry of the International Society for Heart and Lung Transplantation: 29th adult lung and heart-lung transplant report-2012. J Heart Lung Transplant. 2012;31(10):1073–86.
31. Mardock AL, et al. Impact of donor diabetes on outcomes of lung transplantation in the United States. J Surg Res. 2019;244:146–52.
32. Ambur V, et al. The impact of lungs from diabetic donors on lung transplant recipientsdagger. Eur J Cardiothorac Surg. 2017;51(2):285–90.
33. Montori VM, et al. Posttransplantation diabetes: a systematic review of the literature. Diabetes Care. 2002;25(3):583–92.
34. Bando K, et al. Analysis of time-dependent risks for infection, rejection, and death after pulmonary transplantation. J Thorac Cardiovasc Surg. 1995;109(1):49–57; discussion 57–9.
35. Ahmad O, et al. Impact of donor lung pathogenic bacteria on patient outcomes in the immediate post-transplant period. Transpl Infect Dis. 2018;20(6):e12986.
36. Ruiz I, et al. Donor-to-host transmission of bacterial and fungal infections in lung transplantation. Am J Transplant. 2006;6(1):178–82.

Ex Vivo Lung Perfusion

2

Abstract

Ex vivo lung perfusion (EVLP) allows for the evaluation and rehabilitation of donor lungs prior to transplantation. EVLP employs acellular perfusate, lung-protective ventilation, and controlled normothermia to assess an optimized graft function outside the body. EVLP can be performed using different techniques, including the Lund protocol, the Toronto protocol, and the Organ Care System (OCS) protocol. The Toronto protocol emphasizes gradual rewarming and perfusion, with real-time physiological assessment determining suitability for transplant. EVLP typically runs for 3–4 h, allowing safe extension of preservation time. Long-term studies show no difference in survival, graft function, or quality of life between EVLP and non-EVLP recipients, thereby safely expanding the donor pool without compromising outcomes.

2.1 What Is Ex Vivo Lung Perfusion (EVLP)?

EVLP is a technique developed for the assessment, preservation, and rehabilitation of marginal donor lungs [1]. During EVLP, the pulmonary vessels are connected to a circuit through which a centrifugal pump drives acellular perfusate in a temperature-controlled environment (Figs. 2.1 and 2.2). The trachea is simultaneously intubated and connected to a ventilator for lung protective ventilation (PEEP 5, Vt 7 ml/kg, RR 7/min, Fio2 21%). Assessments are performed with PEEP 5, Vt 10 ml/kg, RR 10/min, FiO2 100%.

EVLP can be performed using different techniques, including the Lund protocol, the Toronto protocol, and the Organ Care System (OCS) protocol. Key differences include an open vs. closed left atrium, cross-matched blood vs acellular perfusate, targeted vascular flow as a percentage of donor cardiac output, and centrifugal vs roller pumps. The essential principles of the Toronto EVLP protocol are gradual

M. Salvatori, A. Huang (eds.), *Anesthesia for Lung Transplant*,
https://doi.org/10.1007/978-3-032-06112-6_2

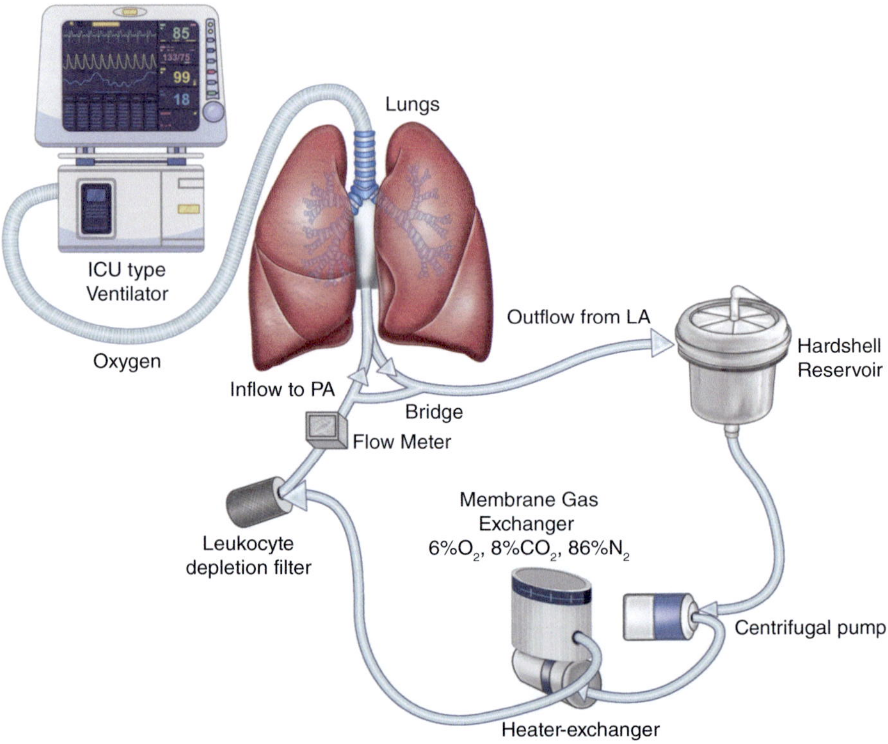

Fig. 2.1 EVLP circuit [4] (ICU, intensive care unit; LA, left atrium; PA, pulmonary artery. *Anesthesiology Clin* 2019; 37:639–660)

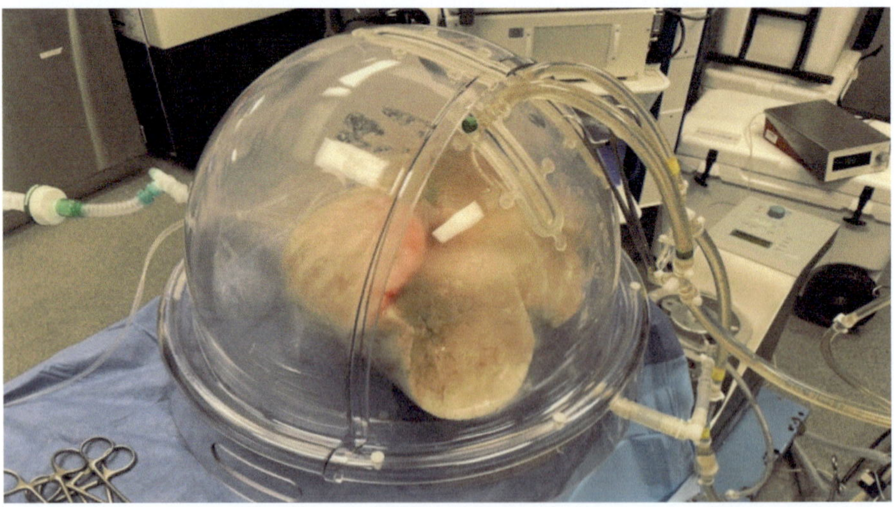

Fig. 2.2 EVLP assessment

Table 2.1 The principles of the Toronto EVLP protocol

- Gradual rewarming of the grafts to normothermia
- Gradual increase in vascular flow as the lung is rewarmed, targeting 40% of the donor cardiac output, to avoid vascular injury/stress and pulmonary edema
- Lung protective ventilation
- Closed left atrium
- Acellular perfusate with increased osmotic pressure to minimize hemolysis

Table 2.2 Indications for EVLP [2]

- PaO2/FiO2 < 300 mmHg
- Signs of pulmonary edema either on chest x-ray or physical examination
- Poor lung compliance on examination at procurement
- High-risk history, such as >10 units of blood transfusion or possible aspiration
- DCDs with >60 min from withdrawal of life support to cardiac arrest

rewarming, gradual increase in vascular flow to 40% of the donor cardiac output, a closed left atrium (LA), and acellular perfusate (Table 2.1). See Machucha et al. (2014) for a concise review of the Toronto EVLP process [2].

Lungs are assessed hourly during EVLP. Pulmonary artery pressure (PAP), left atrial pressure (LAP), peak airway pressure, plateau pressure, and lung compliance are measured. Venous and arterial blood gases are performed. Lung x-ray is performed at 1 h after EVLP start, then every 2 h thereafter. Acceptance criteria include a PF ratio > 400 mmHg and stable or improving PA pressures, airway pressures, and pulmonary compliance. Rejection criteria include a PF ratio < 400 mmHg or a 15% deterioration in PA pressure, airway pressures, or compliance. A decision can typically be made at 3 h, but EVLP time can be extended up to 6 h if necessary. See Yeung et al. [3] for a full description of the physiological assessment of the ex vivo donor lung for transplantation [3].

2.2 What Determines Whether a Lung Requires EVLP?

EVLP rehabilitates lungs that fail criteria for immediate transplantation, including lungs that deviate from one or more "ideal donor" criteria (Table 1.3). Machuca et al. outlined indications for EVLP in DCD and DBD donors [2] (Table 2.2). EVLP use is higher for DCD lungs, which are more likely to deviate from ideal criteria. In fact, between 2008 and 2010, the TLTP used EVLP on all DCD donors. However, this has generally relaxed in the last few years, reflecting growing surgeon comfort with direct-to-transplant DCD grafts. Presently, if a DCD donor meets ideal donor criteria (see Table 1.3), and the time from withdrawal of life support to cardiac arrest is less than 60 min, surgeons may proceed direct to transplant [5].

2.3 Do DBD Lungs Ever Require EVLP?

Yes, DBD lungs frequently require EVLP. Due to critical illness, DBD donors may develop ventilator-acquired pneumonia (VAP), pulmonary edema from neurogenic or non-neurogenic causes, aspiration, and pulmonary emboli, all of which may improve with EVLP.

The TLTP published a recent review of 372 EVLP procedures performed between 2008 and 2017 [6]. In this sample, 53.4% of cases warranting EVLP were high-risk DBD lungs (low PaO2 < 300 mmHg, deteriorating donor PF ratio, abnormal bronchoscopy, aspiration or pneumonia, pulmonary edema, abnormal CXR, pulmonary emboli, or high-risk donor history).

2.4 What Is the Average Duration of EVLP?

The average length of EVLP is 3–4 h. This allows for adequate serial assessment and provides the most valuable information to guide decision-making [3].

2.5 Does EVLP Affect Anesthetic Management?

The anesthetic implications of EVLP during subsequent transplantation have not been reported. However, studies examining the effects of EVLP on basic lung physiology may yield potential therapeutic implications. For example, although pulmonary edema is an indication for EVLP, the process itself may increase total lung water [7, 8]. In a small series of 21 grafts, lung weight increased by 7% after 2 h of EVLP, with weight being negatively correlated with transplant suitability [7, 8]. Although the practical significance of these findings is unclear, EVLP lungs may require additional PEEP to overcome the effects of excess extravascular lung water on lung compliance.

Secondly, animal models show that EVLP enhances arginase activity, which reduces nitric oxide levels in donor lungs [9]. Inhaled nitric oxide (iNO) may therefore be particularly effective in treating post-reperfusion hypoxia or pulmonary hypertension in EVLP lungs.

2.6 Does EVLP Impact Outcomes?

Since its inception in 2008, EVLP has increased the donor pool by approximately 24.6%, 69% of lungs treated with EVLP proceed to transplant [5].

A recent publication from the TLTP reviewed the long-term outcomes of EVLP using the Toronto Protocol [5]. This review compared 706 non-EVLP against 230 EVLP transplants between 2008 and 2017. There were no differences in time to death or retransplant between groups. Furthermore, time to development of chronic allograft dysfunction was comparable. A smaller review of TLTP functional outcomes and quality of life (QoL) showed no significant differences in FEV_1, 6-min walk test or episodes of acute rejection between EVLP and non-EVLP transplants [10]. Furthermore, patients in both groups reported significant improvements in quality of life, with no significant differences between groups. Overall, EVLP does not appear to impact long-term survival, graft function, or patient QoL.

References

1. Cypel M, et al. Normothermic ex vivo lung perfusion in clinical lung transplantation. N Engl J Med. 2011;364(15):1431–40.
2. Machuca TN, Cypel M. Ex vivo lung perfusion. J Thorac Dis. 2014;6(8):1054–62.
3. Yeung JC, et al. Physiologic assessment of the ex vivo donor lung for transplantation. J Heart Lung Transplant. 2012;31(10):1120–6.
4. Lung Transplant Plasma Exchange Protocol for Patients on ECLS, Intraoperative ECMO and Cardiopulmonary Bypass. 2016. https://intranet.remote.uhn.ca/cvpn/zS4DsKk03kE6uu4bm-c8l0FtVxhhCq5M/sites/uhn/ECLS/Plasmapheresis%20Guideline%202016.pdf.
5. Divithotawela C, et al. Long-term outcomes of lung transplant with ex vivo lung perfusion. JAMA Surg. 2019;154(12):1143–50.
6. Cypel M, et al. Normothermic ex vivo lung perfusion: does the indication impact organ utilization and patient outcomes after transplantation? J Thorac Cardiovasc Surg. 2019;159(1):346–55.e1.
7. Okamoto T, et al. Donor lung weight at lung procurement; predictive value for transplant suitability during ex vivo lung perfusion. J Heart Lung Transplant. 2019;38(4):S188.
8. Okamoto T, et al. Significance of lung weight in cellular ex vivo lung perfusion. J Surg Res. 2021;260:190–9.
9. Tavasoli F, et al. Increased arginase expression and decreased nitric oxide in pig donor lungs after Normothermic ex vivo lung perfusion. Biomolecules 2020;10(2).
10. Tikkanen JM, et al. Functional outcomes and quality of life after normothermic ex vivo lung perfusion lung transplantation. J Heart Lung Transplant. 2015;34(4):547–56.

3 Recipients

Abstract

Lung transplant candidacy depends on multidisciplinary evaluation of absolute and relative contraindications, including advanced age, comorbidities, and infection risk, all of which influence post-transplant outcomes. Comprehensive pre-listing investigations assess organ function, cardiovascular status, and psychosocial readiness. Anesthetic consultation reviews perioperative risks including graft dysfunction, surgical complications, extracorporeal support, and intensive care. Rigorous evaluation and optimization improves candidate selection and postoperative outcomes.

3.1 What Are the Most Common Indications for Transplant at TGH? (Table 3.1)

Table 3.1 Indications for lung transplantation at TGH (2017–2021)

Indication	%
Interstitial disease	51
Obstructive disease	31
Cystic fibrosis (CF)	9
Re-transplantation/graft failure	5
Primary pulmonary hypertension (PHTN)	3
Other	2

M. Salvatori, A. Huang (eds.), *Anesthesia for Lung Transplant*,
https://doi.org/10.1007/978-3-032-06112-6_3

3.2 Which Recipient Features Increase Risk?

The ISHLT's consensus document for the selection of transplant candidates outlines risk factors that preclude transplantation (absolute contraindications), as well as those that are associated with increased risk (but may still be listed and transplanted) [1].

Candidates with absolute contraindications are considered too high risk to achieve successful outcomes post lung transplantation (Table 3.2). Most lung transplant programs should not transplant patients with these risk factors except under very exceptional or extenuating circumstances.

Candidates with high or substantially increased risk (Table 3.3) may be considered in centres with expertise specific to the condition. Data supporting transplantation under these circumstances may be limited or show substantially increased risk. When 2 or more risk factors are present, they compound the risk of adverse outcomes. Modifiable conditions should be optimized when possible.

In addition, there are several risk factors with unfavourable short- and long-term outcomes after lung transplant (Tables 3.4). Multiple risk factors together may increase the risk for adverse post-transplant outcomes.

Table 3.2 Absolute contraindications to transplantation[2]

- Lack of patient willingness or acceptance
- Malignancy with high risk of recurrence or death related to cancer
- Glomerular filtration rate (GFR) < 40 mL/min/1.73^{m2} unless being considered for multi-organ transplant
- Acute coronary syndrome or myocardial infarction within 30 days (excluding demand ischemia)
- Stroke within 30 days
- Liver cirrhosis with portal hypertension or synthetic dysfunction unless being considered for multi-organ transplant
- Acute liver failure
- Acute renal failure with rising creatinine or on dialysis and low likelihood of recovery
- Septic shock
- Active extrapulmonary or disseminated infection
- Active tuberculosis infection
- Human Immunodeficiency Virus (HIV) infection with detectable viral load
- Limited functional status with poor potential for post-transplant rehabilitation
- Progressive cognitive impairment
- Repeated episodes of non-adherence without evidence of improvement (adults)
- Active substance use or dependence including current tobacco use, vaping, marijuana smoking, or intravenous (IV) drug use
- Other severe medical conditions expected to limit survival after transplant

Table 3.3 Candidates with substantially increased risk [1]

- Age > 70 years
- Severe coronary artery disease (CAD) that requires coronary artery bypass grafting (CABG) at transplant
- Reduced left ventricular ejection fraction (LVEF) <40%
- Significant cerebrovascular disease
- Severe esophageal dysmotility
- Untreatable hematologic disorders (bleeding diathesis, thrombophilia, severe bone marrow dysfunction)
- Body Mass Index (BMI) > 35 or < 16
- Limited functional status with potential for post-transplant rehabilitation
- Unreliable support system or caregiving plan
- Lack of understanding of disease and/or transplant despite teaching
- Specific infections:
  - *Mycobacterium abscessus*
  - *Lomentospora prolificans*
  - *Burkholderia cenocepacia* or *gladioli*
- Hepatitis B or C infection with detectable viral loads and liver fibrosis
- Chest wall or spinal deformity expected to cause restriction after transplant
- Extracorporeal life support (ECLS)
- Retransplant:
  - Within a year of initial lung transplant
  - Due to restrictive allograft syndrome (RAS) subtype of Chronic Lung Allograft Dysfunction (CLAD)
  - Due to antibody-mediated rejection as etiology for CLAD

Table 3.4 Factors which may compromise outcomes [1]

- Age 65–70 years
- GFR 40–60 mL/min/1.73m
- Mild to moderate coronary artery disease
- Severe coronary artery disease that can be revascularized via PCI (percutaneous coronary intervention) prior to transplant
- Patients with prior coronary artery bypass grafting
- Reduced left ventricular ejection fraction (LVEF) 40–50%
- Peripheral vascular disease (PVD)
- Connective tissue diseases (CTD)
- Severe gastroesophageal reflux disease (GERD)
- Esophageal dysmotility
- Thrombocytopenia, leukopenia, or anemia with high likelihood of persistence post-transplant
- Osteoporosis
- BMI between 30–34.9 or between 16–17
- Frailty
- Hypoalbuminemia
- Poorly controlled diabetes
- Edible cannabis use
- *Scedosporium apiospermum* infection
- HIV infection with undetectable viral loads
- Previous thoracic surgery
- Prior pleurodesis
- Mechanical ventilation
- Retransplant >1 year for obstructive CLAD

3.3 What Investigations Occur with the Listing Assessment? (Table 3.5)

Table 3.5 Investigations performed during listing assessment

Routine investigations for all recipients
• Routine bloodwork: complete blood count (CBC), electrolytes, creatinine, liver enzymes • Human Leukocyte Antigen (HLA) testing: Initially + every 3 months while on wait list • Serologies: Epstein–Barr virus (EBV), Cytomegalovirus (CMV), hepatitis B and C, HIV • Cardiac testing: Transthoracic echocardiogram • Chest imaging/studies: Chest x-ray (CXR), Computed Tomography (CT) scan of the chest, and Ventilation/Perfusion (V/Q) study • Pulmonary testing: Pulmonary Function Tests (PFTs) and 6-Minute Walk Test (6MWT) • Cancer screening: Fecal occult blood, pap smear, and mammograms
Diagnosis-specific investigations
• Rheumatoid arthritis: Flexion/extension views of c-spine • Cystic fibrosis (CF) patients: Abdominal ultrasound and CT of sinuses • Scleroderma: Gastrointestinal (GI) studies (barium swallow, esophageal manometry) • Obese and/or alcohol history: Abdominal ultrasound
Additional consultations
• Social work assessment • Dietician assessment: TLTP has a BMI < 30 requirement for listing • Psychology assessment: If known mental health issues, or concerns raised during assessment • Transitional pain service (TPS) assessment: In patients with known chronic pain or significant pre-transplant opioid use (goal, if listed, is to wean to less than 100 mg of oral morphine equivalents per day)
Additional cardiovascular testing
• Pharmacologic stress test (persantine) for all patients over 40 • Carotid dopplers for patients over 50, or those with extensive smoking history or known vascular disease • Right heart catheterization (RHC) for patients with abnormal echo, advanced compatible disease (i.e., Ideopathic Pulmonary Fibrosis (IPF), connective tissue disorder), or clear primary diagnosis (primary pulmonary hypertension) • Left heart catheterization for positive stress test, history of cardiac disease, or multiple cardiac risk factors • Referral to UHN cardiologist and/or cardiac surgeon for consultation if any cardiac investigations are abnormal

3.4 When Is Single-Lung Transplant Considered?

In general, single lung transplantation (SLT) is reserved for patients with pulmonary fibrosis without pulmonary hypertension over the age of 65, as they show equivalent survival to double lung transplants (DLTs) [2]. SLT for COPD varies by centre, but the TLTP favours DLT for COPD, owing to possible survival benefit and hyperinflation of the native lung compromising the function of the graft [3]. Chronic infection and colonization also preclude patients from SLT. This includes patients with septic lung disease such as cystic fibrosis and bronchiectasis. Additionally, patients with other lung diseases colonized with microorganisms would also be excluded from SLT consideration. Candidates with primary pulmonary arterial hypertension also require DLT due to bilateral involvement of the pulmonary vasculature. Table 3.6 shows differences in spirometry for SLT vs DLT [4].

Traditionally, SLTs were performed on the side receiving lesser perfusion to facilitate oxygenation during the surgery. However, as transplant on Extracorporeal Membrane Oxygenation (ECMO) becomes routine, sidedness is increasingly determined by graft quality.

For patients with small thoracic cavities, lobar transplantation reduces the wait time for appropriately sized grafts. However, lobar grafts are particularly susceptible to reperfusion injury due to their reduced pulmonary vasculature compared to a whole lung of the same size, necessitating intraoperative ECMO and frequent need for post-operative ECMO.

Table 3.6 Percent predicted Forced Expiratory Volume in 1 Second (FEV1) after SLT versus DLT [4]

Timepoint	SLT	DLT
Post-transplant	50	60
1 year	55	75
3 years	47	65

[a]In both groups, lower FEV_1 was associated with increased mortality, but this effect was more pronounced in the SLT group

3.5 Which Major Perioperative Risks Should Be Discussed with All Patients During Consultation? (Table 3.7)

Table 3.7 Risks discussed during consultation

- General description of anesthesia and associated risks—Dental/oral trauma, invasive arterial and venous access, etc.
- Risk of requiring ECLS and the potential for increased risk of bleeding, stroke, myocardial infarction (MI), and death
- Risks associated with blood transfusion
- Risks of primary graft dysfunction and early rejection
- Anastomotic complications: Airway leaks, pulmonary artery stenosis, kinking of pulmonary artery, and pulmonary vein thrombosis
- Phrenic nerve injury and diaphragmatic dysfunction—Up to 30% [5]
- Potential for prolonged ICU stay, mechanical ventilation, and potential tracheostomy

References

1. Leard LE, et al. Consensus document for the selection of lung transplant candidates: an update from the International Society for Heart and Lung Transplantation. J Heart Lung Transplant. 2021;40(11):1349–79.
2. Ranganath NK, et al. Single and double lung transplantation have equivalent survival for idiopathic pulmonary fibrosis. Ann Thorac Surg. 2020;109(1):211–7.
3. Crawford TC, et al. Five-year mortality Hazard is reduced in chronic obstructive pulmonary disease patients receiving double- versus single-lung transplants. J Surg Res. 2019;237:118–25.
4. Mason DP, et al. Effect of changes in postoperative spirometry on survival after lung transplantation. J Thorac Cardiovasc Surg. 2012;144(1):197–203.
5. Sheridan PH Jr, et al. Incidence of phrenic neuropathy after isolated lung transplantation. The Loyola university lung transplant group. J Heart Lung Transplant. 1995;14(4):684–91.

Preoperative Planning and Monitors

4

Abstract

Lung transplantation presents unique perioperative challenges for the anesthesiologist. Preoperative reassessment is essential, as candidates may have deteriorated since their initial evaluation. Anesthetic management must be tailored to accommodate complex comorbidities, often requiring specialized equipment and medications. Sensitized recipients may require intraoperative plasma exchange to mitigate antibody-mediated rejection. Optimal outcomes rely on effective perioperative collaboration among the anesthesiology, surgery, perfusion, and critical care teams.

4.1 How Is the Room Setup for a Lung Transplant? (Table 4.1)

Table 4.1 Standard lung transplant checklist

Equipment	
Anesthesia machine	• Vaporizer filled • Circuit checked • Working suction • Bag valve mask (BMV) • Positive End Expiratory Pressure (PEEP) valve
Airway equipment	• Oropharyngeal airway • Laryngoscope or video laryngoscope • Double-lumen endotracheal tubes (DLETTs), various sizes • Cook extra firm soft tip airway exchange catheter, 11Fr (size 35 DLETT) or 14Fr, 100 cm (all other DLETT sizes) • ETT with supra-glottic suction (Ventilator Associated Pneumonia [VAP] tube) • Bronchoscopy tower • Pediatric and adult bronchoscopes

(continued)

M. Salvatori, A. Huang (eds.), *Anesthesia for Lung Transplant*,
https://doi.org/10.1007/978-3-032-06112-6_4

Table 4.1 (continued)

Additional equipment	• Large-bore peripheral intravenous cannula (IV) • Radial arterial line • Central line kit (Cordis 8.5Fr) + Swan-Ganz catheter • Triple-lumen central line • Ultrasound for central line placement • 3-in-1 pressure transducer setup • Depth of anesthesia monitor • Naso/oropharyngeal temperature probe • Fluid warmer (if high risk of transfusion) • Underbody warmer
Drugs	
Induction agents and analgesia	• Midazolam (0.01–0.05 mg/kg) • Fentanyl (max 10 μg/kg) • Propofol (0.5–2 mg/kg) • +/− Ketamine (0.5–1 mg/kg) • Rocuronium (0.6–1.2 mg/kg) • Hydromorphone (0.01–0.03 mg/kg) • +/− Sufentanil (10–30 μg/hr)
Rescue meds	• Phenylephrine (premixed bags, 40 μg/ml) • Ephedrine (dilute 50 mg/10 mls) • Epinephrine (premixed bags, 4 μg/ml) for bolus • Calcium chloride (1 g/10 mls)
Infusions	• Norepinephrine (premixed bags, 32 μg/ml) • Vasopressin (dilute 40 u/100 mls) • Propofol (10 mg/ml) • Normal saline 500 cc as driveline w/ 5 g magnesium sulfate ($MgSO_4$)
Antibiotics and steroids	• Piperacillin/Tazobactam (4.5 g) q6h • Solumedrol (methylprednisolone) 500 mg
Miscellaneous	• Heparin (400–600 IU/kg in case of emergency cardiopulmonary bypass (CPB)) • Tranexamic acid (up to 50 mg/kg or 5 g) • Antiarrhythmics: Esmolol, metoprolol, amiodarone, $MgSO_4$, lidocaine • Salbutamol

4.2 Are There Any Recipients that Require a Modified Setup? (Table 4.2)

Table 4.2 Modifications for Specific Indications

Indication	Modifications
Chronic Obstructive Pulmonary Disease (COPD)	• Bronchodilators prior to induction if reversibility seen on PFTs • Ketamine/magnesium/volatiles for bronchodilatation
Severe pulmonary hypertension (PHTN)	• Consider awake lines • Defibrillator pads • Patient prepped and draped, surgeons scrubbed • +/− wires/cannulae in femoral vessels • Transesophageal echo (TEE) to assess right ventricle (RV) function
Cystic fibrosis (CF)	• Additional antibiotics as guided by Transplant Infectious Disease Team • Thorough bronchopulmonary toilet • +/− chest and DLETT povidone washout
Redo transplant	• Blood in room from case start • Fluid warmers, wide-bore IV access
Combined heart-lung transplant	• Defib pads • TEE • Higher steroid dose • Heparin 400–600 IU/kg

4.3 How Does the Patient Arrive to the Operating Room (OR)?

Once viability of donor lungs is confirmed, patients will come to the OR either from home via the ward or from the Intensive Care Unit (ICU). Stable ward patients will be brought to the hallway outside of the OR for nursing check-in, surgical consent, and final anesthetic assessment. Anesthesia personnel must transport any patient from an ICU directly into the OR with full monitoring. Patients on ECMO also require a respiratory therapist (RT) and perfusionist for transport.

4.4 How Is the Patient Assessed Preoperatively?

Most transplant candidates have had an extensive workup including anesthetic consultation. However, the time elapsed between consult/investigations and transplant is extremely variable, and clinical status may deteriorate in the interim. The interstitial lung disease (ILD) population are particularly susceptible to rapid deteriorations. The purpose of the final anesthetic assessment is to assess for changes that may modify your plan.

Begin with a global assessment of the patient, focusing on comfort at rest, level of respiratory effort, dyspnea, distress, and level of supplemental oxygen required. Note any discrepancies from the time of consultation. When possible, delineate the time course of deterioration in relation to prior investigations. If the patient has become more symptomatic since the last echo or right heart catheterization, recognize that the pulmonary artery pressures may have also risen in the interim. A brief review of systems should identify whether there have been any major changes in other organ systems. As many of the consultations now occur via telemedicine, predictors of difficult bag mask ventilation (BMV) or intubation may not have been identified. The airway assessment for lung transplant recipients must be thorough and accurate, due to the rapidity of desaturation, low compliance, inadequacy of BMV, and the size of the double-lumen ETT.

Check ongoing infusions as steroids or antibiotics may have been started on the ward. For hospitalized patients, review the 7-day medical administration record (MAR), looking specifically for anticoagulation and diuresis. Be prepared to reverse anticoagulation when necessary. Recent aggressive diuresis may result in hypotension unresponsive to pressors which further complicates induction, especially in the context of high pulmonary artery (PA) pressures.

4.5 How Should the Patient Be Transferred and Positioned?

Once the patient is in the OR, we strongly recommend that all patients be transferred passively using a sliding board, even if independently mobile. The exertion associated with independent transfer can trigger a prolonged desaturation and dyspnea that complicates induction and/or awake lines.

Remember that many recipients will be unable to tolerate lying flat without significant discomfort or dyspnea. Arterial line, peripheral IV, and preoxygenation should all be performed in a semi-recumbent position with the head elevated to approximately 30 degrees. If awake lines are required, slowly lower the back of the patient to the lowest tolerable position. Scan the internal jugular vein before scrubbing, as the vein may be too small and collapsible, especially in the context of aggressive preoperative diuresis or gasping breaths. If awake lines are required and the patient cannot tolerate lying flat, consider alternative such as infusing dilute vasopressor solutions through wide-bore peripheral IVs, or inserting a pre-induction femoral line, which can be performed safely in the reverse Trendelenburg position.

4.6 Is there a Role for Cerebral Oximetry During Lung Transplant?

Cerebral oximetry using near infrared spectroscopy (NIRS) is not routine during lung transplantation, with only 41% of centers reporting regular use [1]. However, the rates of delirium and cognitive dysfunction following transplant are underappreciated, with estimates as high as 40 and 60% respectively [1–3]. Chronic hypoxemia is thought to be a major contributor, as many recipients show evidence of cognitive impairment even prior to transplantation, with home oxygen bestowing some protection [4]. However, Tomasi et al. (2021) showed that operative time is the only independent risk factor for postoperative cognitive dysfunction after lung transplant [1]. Notably, no preoperative risk factors were identified, and intraoperative cerebral desaturation on NIRS was not predictive of neurocognitive outcomes.

4.7 When Should Transesophageal Echocardiography (TEE) be Used?

TEE use in lung transplantation is not accepted universally and remains a class IIb recommendation by the American Society of Echocardiography (*Class IIb recommendation: Usefulness/efficacy is less well established by evidence/opinion*) [5]. At TGH, TEE use varies by staff, clinical circumstances, and availability. Indications for TEE during lung transplantation are listed in Table 4.3. The key features of the TEE assessment for lung transplantation are listed in Table 4.4.

There have been no randomized controlled trials that demonstrate improved outcomes when TEE is used for lung transplantation. However, there are several case reports that demonstrate the value of TEE in the diagnosis and management of

Table 4.3 Indications for TEE during lung transplant

- Moderate to severe LV or RV dysfunction
- Known valvular pathology
- Severe pulmonary hypertension
- Combined lung transplant/CABG or heart-lung transplant
- Inability to achieve adequate ECMO flows
- Unexplained hemodynamic instability or refractory hypoxemia
- Donor-recipient size mismatch
- Surgical concern regarding any of the vascular anastomoses

Table 4.4 Key features of TEE for lung transplant

Baseline	Post-transplant
• RV size and function • Valve issues (Mitral Stenosis (MS), Mitral Regurgitation (MR), Tricuspid Regurgitation (TR) • LV function • Patent Foramen Ovale (PFO) assessment • Baseline pulmonary vein assessment	• PA anastomosis *Stenosis indicated by:*[10] – Mean velocity > 2.6 m/s – Peak pressure gradient >57 mmHg – Branch PA diameter < 0.8 cm – ≥ 25% narrowing from native PA – Turbulent flow on color doppler – Presence of thrombus • PV anastomoses *Stenosis indicated by* [11, 12]*:* – Mean velocity > 1 m/s – PV diameter < 0.5 cm • RV size and function – RV dysfunction if FAC <35%, TAPSE <1.7 cm • Volume status • PFO reassessment • Guide peripheral ECMO cannulation if required

anastomotic complications and hemodynamic instability during lung transplant surgery [6–9].

It can be very challenging to setup the echo machine in an emergency. If you anticipate any cardiovascular instability, call the echo team early to prepare (ideally pre-induction), even if TEE is not ultimately required.

4.8 Is there a Role for Continuous Cardiac Output (CCO) PA Catheters During Lung Transplant?

The standard PA catheters used at TGH are Intermittent Cardiac Output (ICO) Swan-Ganz catheters. Advanced hemodynamic monitoring is possible with Continuous Cardiac Output (CCO) Swan-Ganz catheters, also known as "Oximetric Swans". They afford several advantages, including continuous measurement of cardiac output (CO), mixed venous oxygen saturation (SvO2), end diastolic volume (EDV), and right ventricular ejection fraction (RVEF) without the need to peform repeated bolus thermodilution or sample venous gases. These are the PA catheters typically used during pulmonary endarterectomies, when limiting cardiac output is essential for preventing reperfusion injury.

There have been no studies comparing lung transplant outcomes in patients with ICO vs. CCO PA catheters. Theoretically, CCO PA catheters can help manage patients with known cardiac dysfunction when TEE is not available. Ischemic-reperfusion injury post transplantation is a complex topic and limiting cardiac output has not been shown to benefit long-term graft function. This is likely because the ischemia in donor lungs is acute, and the pulmonary vasculature has not adapted to a hypoxemic, low-flow state, as with CTEPH.

4.9 Which Patients Need Plasma Exchange (PLEX) and What Are the Implications?

Intraoperative plasma exchange (PLEX) is used when recipients express pre-formed antibodies to proteins in the donor lungs. These donor HLA-specific antibodies (DSAs) can cause hyperacute rejection, antibody-mediated rejection (AMR), and early onset chronic lung allograft dysfunction (CLAD) with poor outcomes after lung transplantation. Perioperative plasma exchange aims to remove these DSA from the recipient plasma, thereby reducing the risk of rejection. The TLTP has published short [13] and long-term [14] outcomes for DSA-positive recipients treated with a perioperative desensitization protocol involving perioperative plasma exchange, intravenous immune globulin (IVIG), antithymocyte antiglobulin (ATG), +/− mycophenolic acid (MPA). In brief, there were no differences in primary graft dysfunction, allograft survival, or CLAD-free survival between unsensitized patients and DSA-positive patients treated with perioperative desensitization [13]. A subsequent follow-up study shows that 5- and 10-year mortality is also equivalent between groups [14].

The anesthetic implications of the UHN Lung Transplant Plasma Exchange Protocol for Patients on ECLS, Intraoperative ECMO, and Cardiopulmonary Bypass are summarized below:

Access. PLEX requires an independent circuit and cannot be performed via ECMO or CPB. The transplant surgeon places a femoral hemodialysis catheter at the start of the surgery. The hemodialysis catheter remains in place for post-transplant PLEX, which may continue for weeks.

Anticoagulation. In addition to an apheresis nurse, a perfusionist must be present for the duration of PLEX to document the procedure and monitor systemic anticoagulation. A heparin infusion is prepared by the perfusionist and administered via the apheresis return line at 2000 IU/hr for transplants performed on CPB or ECMO. As usual, the activated clotting time (ACT) targets are 160–180 s on ECMO, or 480–600 s on CPB. ACT levels are drawn every 15–30 mins from the CPB circuit or arterial line. If not within target range, the perfusionist will titrate the heparin infusion accordingly.

Fluid Management. The apheresis nurse will calculate the volume of fluid replacement based on the patient's height, weight, and hematocrit. Three plasma volumes will be processed. The first two plasma volumes are replaced with 5% albumin. The third plasma volume is replaced with fresh frozen plasma (FFP). The apheresis nurse will notify the perfusionist 15 min before switching to FFP, as the ACT commonly lowers.

Calcium. The apheresis nurse runs a calcium infusion throughout PLEX to offset calcium-citrate chelation. Perform regular ABGs and communicate the calcium level to the apheresis nurse so that dose adjustments can be made.

Acidosis. Albumin is a weak acid that can cause metabolic acidosis in large volumes [15]. Lab values typically show low bicarbonate (HCO3) with normal lactate and chloride [16]. A sodium bicarbonate (NaHCO3) infusion may be required to maintain a normal pH.

Duration. On average, the duration of PLEX is 2–4 h depending on the patient's plasma volume. Ideally, intraoperative PLEX should be complete before reperfusion of the first lung. Reversal of heparin anticoagulation following PLEX is at the discretion of the surgeon.

4.10 What Modifications Are Made for Dialysis-Dependent Patients?

This is a rare occurrence, as Ontario's Referral and Listing Criteria for Adult Lung Transplantation (Version 3.0) indicates that "patients with untreatable advanced dysfunction of another major organ system (e.g., heart, liver, or kidney) may not be eligible for lung transplantation unless combined organ transplantation can be performed" [17].

Should a dialysis-dependent patient require lung transplantation, it will likely be sufficient to dialyze immediately pre- and post-surgery. Although intraoperative dialysis is common during heart surgery, the same is not true of lung transplants as the patient remains normothermic, there is no potassium-containing cardioplegia, primary metabolic disturbances are rare, and coagulopathy and massive transfusion are unlikely.

In rare cases where intraoperative dialysis is indicated, patients will require a designated hemodialysis catheter. Remember to account for fluid shifts, electrolyte changes, and circuit heparin in close communication with the nephrology team. Hemodialysis cannot be performed via the ECMO circuit. Patients on CPB (combined lung-CABG or lung-heart transplant) can be easily dialyzed by the perfusionist while on bypass without the involvement of the nephrology team.

References

1. Tomasi R, et al. Intraoperative anesthetic Management of Lung Transplantation: center-specific practices and geographic and centers size differences. J Cardiothorac Vasc Anesth. 2018;32(1):62–9.
2. Smith PJ, et al. Neurocognitive changes after lung transplantation. Ann Am Thorac Soc. 2014;11(10):1520–7.
3. Cohen DG, et al. Cognitive function, mental health, and health-related quality of life after lung transplantation. Ann Am Thorac Soc. 2014;11(4):522–30.
4. Thakur N, et al. COPD and cognitive impairment: the role of hypoxemia and oxygen therapy. Int J Chron Obstruct Pulmon Dis. 2010;5:263–9.
5. Cheitlin MD, et al. ACC/AHA/ASE 2003 guideline update for the clinical application of echocardiography: summary article. A report of the American College of Cardiology/American Heart Association task force on practice guidelines (ACC/AHA/ASE Committee to update the 1997 guidelines for the clinical application of echocardiography). J Am Soc Echocardiogr. 2003;16(10):1091–110.
6. Abrams BA, et al. TEE for lung transplantation: a case series and discussion of vascular complications. J Cardiothorac Vasc Anesth. 2020;34(3):733–40.
7. Felten ML, et al. Transesophageal and contact ultrasound echographic assessments of pulmonary vessels in bilateral lung transplantation. Ann Thorac Surg. 2012;93(4):1094–100.
8. Huang YC, et al. Graft failure caused by pulmonary venous obstruction diagnosed by intraoperative transesophageal echocardiography during lung transplantation. Anesth Analg. 2000;91(3):558–60.
9. Gonzalez-Fernandez C, et al. Pulmonary venous obstruction after lung transplantation. Diagnostic advantages of transesophageal echocardiography. Clin Transpl. 2009;23(6):975–80.
10. Kumar N, et al. Evaluating the impact of pulmonary artery obstruction after lung transplant surgery: a systematic review and meta-analysis. Transplantation. 2021;105(4):711–22.
11. Michel-Cherqui M, et al. Intraoperative transesophageal echocardiographic assessment of vascular anastomoses in lung transplantation. A report on 18 cases. Chest. 1997;111(5):1229–35.
12. Kumar N, et al. Pulmonary cuff dysfunction after lung transplant surgery: a systematic review of the evidence and analysis of its clinical implications. J Heart Lung Transplant. 2019;38(5):530–44.
13. Tinckam KJ, et al. Survival in sensitized lung transplant recipients with perioperative desensitization. Am J Transplant. 2015;15(2):417–26.
14. Aversa M, et al. Long-term outcomes of sensitized lung transplant recipients after perioperative desensitization. Am J Transplant. 2021;21(10):3444–8.
15. Langer T, et al. Effects of intravenous solutions on acid-base equilibrium: from crystalloids to colloids and blood components. Anaesthesiol Intensive Ther. 2014;46(5):350–60.
16. Fidkowski C, Helstrom J. Diagnosing metabolic acidosis in the critically ill: bridging the anion gap, Stewart, and base excess methods. Can J Anaesth. 2009;56(3):247–56.
17. TGLN. Ontario's Referral and Listing Criteria for Adult Lung Transplantation v3.0. 2018. https://www.giftoflife.on.ca/resources/pdf/ON_Adult_Lung_Transplant_Referral_and_Listing_Criteria_3.0_Aug_2018.pdf.

Induction

5

Abstract

In lung transplant candidates with severe hypoxemia, pulmonary hypertension, or right ventricular (RV) dysfunction, induction of anesthesia poses significant risk. In extreme cases, pre-induction central lines or ECMO cannulation can offer additional cardiopulmonary support. Induction strategies should maintain RV perfusion, avoid hypoxia, hypercapnia, and acidosis, and tailor ventilation to underlying physiology. The anesthesiologist must always be prepared for cardiac arrest on induction, which requires rapid correction of physiological derangements, tailored ACLS, and early extracorporeal support.

5.1 Which Patients Need Awake Central Lines?

Consider pre-induction or awake central lines in patients with marginal reserve that require vasoactive or inotropic support during induction. Awake lines are most common for patients with severe pulmonary hypertension and right heart failure, in which vasopressors are essential for maintaining RV perfusion. Should the patient rapidly deteriorate during induction, awake lines also enable central administration of resuscitative drugs such as epinephrine.

A common indication for awake lines is severe pulmonary hypertension. As discussed in **Question 4.4**, recognize that PA pressures may have risen significantly since the last cath, especially with a recent deterioration in clinic status.

With adequate reassurance and careful positioning, awake lines are feasible for most patients. However, some patients may not tolerate awake lines due to a combination of anxiety, claustrophobia, dyspnea, discomfort, and positioning. For these patients, consider judicious sedation, recognizing that hypercapnia and hypoxia will exacerbate PA pressures. If challenges persist, consider alternatives such as infusing

M. Salvatori, A. Huang (eds.), *Anesthesia for Lung Transplant*,
https://doi.org/10.1007/978-3-032-06112-6_5

dilute vasopressor solutions through wide-bore peripheral IVs, or a femoral line, which can be performed safely in reverse Trendelenburg position (see **Question 4.5**).

5.2 Which Patients Need Awake ECMO Cannulation?

The indications for awake ECMO cannulation are not clearly defined. Discuss awake cannulation with the surgical team for patients at high risk of cardiorespiratory collapse on induction. Examples include patients with suprasystemic PA pressures, RV failure, profound baseline hypoxemia, or prohibitively difficult airway anatomy.

Although pre-induction ECMO is often perceived as a "failsafe", success is not guaranteed due to the challenges of awake cannulation in unstable patients. Femoral cannulation requires immobility and supine positioning, both of which are problematic for the dyspneic patient (see **Question 5.1**). Despite local anesthetic, percutaneous puncture and dilation of groin vessels for the larger ECMO cannulae can be painful and may cause significant bleeding. Additional risks include vessel dissection and retroperitoneal hematoma. The tip of the venous return cannula must be positioned in the right atrium, and optimization can be difficult without TEE guidance. Improper placement of the venous cannula can lead to poor venous drainage, partial ECMO flows, poor RV decompression, and inadequate mechanical support. Finally, recall that systemic perfusion in peripheral veno-arterial (VA)-ECMO depends on retrograde aortic flow which significantly increases LV afterload. In patients with poor LV function, this augmented afterload may be poorly tolerated.

Despite these challenges, awake ECMO cannulation may be lifesaving for brittle patients undergoing induction of anesthesia. If awake cannulation cannot be achieved, insertion of wires/sheaths or surgical exposure of the groin vessels (i.e., "cut downs") may still be possible, and can expedite subsequent cannulation in case of arrest/instability. If full ECMO flows can be achieved, induction and intubation can proceed in a slow and controlled manner, remaining cognizant of the indication for awake ECMO. An induction strategy incorporating benzodiazepines, opioids, and ketamine may minimize vasoplegia and better maintain systemic perfusion pressures.

5.3 What Are the Hemodynamic and Ventilation Targets During Induction?

When possible, maintain the patient's baseline respiratory rate, tidal volume, and breathing pattern to preserve their compensated physiology. Draw a pre-induction arterial blood gas (ABG) to assess baseline oxygenation and carbon dioxide (CO_2) retention. The patient's respiratory disease determines the ideal ventilation strategy.

Patients with obstructive pathologies such as COPD and alpha-1 antitrypsin (A1AT) deficiency will need a longer expiratory phase, and caution must be taken to avoid gas trapping and hyperinflation, especially with bag mask ventilation.

Common starting ventilatory parameters for obstructive disease are tidal volume (Vt) 6-8 cc/kg (~400-450 cc) × 10 breaths per minute (bpm), with an inspiratory:expiratory (I:E) ratio of at least 1:3 (often 1:4 or more). Severe hyperinflation may cause venous compression and loss of venous return resulting in hemodynamic collapse. If hyperinflation is suspected, disconnect the circuit to assist in expiration.

Patients with severe restrictive physiology have reduced total lung capacity (TLC) and high peak inspiratory pressures that benefit from high rates and low tidal volumes. Common starting ventilatory parameters for restrictive disease are Vt 4-6 cc/kg (~350-400 cc) × 16 bpm, with an I:E ratio of 1:2. For patients with mixed obstructive-restrictive disease (i.e., CF, bronchiectasis), vent parameters should be guided by the patient's dominant physiology as demonstrated on pulmonary function tests.

Significant restrictive lung physiology presents challenges for mechanical ventilation. Modern ventilators offer both volume control ventilation (VCV) and pressure control ventilation (PCV) modes, however both these strategies have potential shortcomings in ILD patients with poor lung compliance. VCV may achieve target tidal volumes at the cost of excessive airway pressures. PCV may deliver low tidal volumes for the set pressure, leading to hypoventilation and CO_2 accumulation. Both phenomena (excessive ventilatory pressures or underventilation) can exacerbate underlying pulmonary hypertension, which is common in ILD. If tolerated, aggressive ventilation with high pressures and rates may be necessary to maintain adequate acid-base status. In a double lung transplant, this may be acceptable. However, in single lung transplantation, a native lung will remain, requiring attention to avoid lung injury.

Beware that the anesthesia ventilators at TGH default to a pressure limit of 40 cmH_2O. The ventilators will automatically reduce tidal volumes to satisfy this default pressure limit regardless of the volumes set. Because the pressure limit is not exceeded, the alarm will not sound, and profound hypoventilation may go unrecognized. This can lead to hemodynamic instability and subsequent cardiac arrest in the peri-induction period. Raise the airway pressure alarm limit prior to induction to minimize the risk.

Repeat an ABG within 30 minutes of induction, typically upon completion of central venous access. CO_2 rises rapidly if ventilatory settings are inadequate, and end tidal carbon dioxide ($EtCO_2$) is poorly predictive of the partial pressure of carbon dioxide in blood ($PaCO_2$) due to significant dead space. Worsening hypercapnia despite aggressive ventilation on two lungs should prompt a discussion with the surgical team regarding early ECMO support. For chronic CO_2 retainers, normalize the $PaCO_2$ gradually over hours to allow time for metabolic adaptation.

Regarding hemodynamics, blood pressure (BP) should be maintained within 20% of baseline, generally targeting a mean arterial pressure (MAP) of 60-70 mmHg. In patients with severe pulmonary hypertension or a failing right heart, maintaining the diastolic pressure at or above the patient's baseline is essential to RV perfusion in the context of an elevated right ventricular end diastolic pressure (RVEDP).

5.4 What Are the Standard Antibiotics and Steroids?

Antibiotics: Tazocin (Piperacillin-Tazobactam) 4.5 g IV q6h
Steroids: Methylprednisolone 500 mg IV before skin incision

Confirm that antibiotics and steroids were not given on the ward prior to transfer to avoid repeat dosing. There may be modifications based on allergies or donor microbiology. If there are any questions regarding appropriate antimicrobial therapy, contact the Transplant ID team.

5.5 What Are Typical Induction Drugs and Doses?

Typical induction agents include fentanyl, midazolam, ketamine, and propofol (Table 5.1). Induction drug and dose choices are variable, and should reflect the patient's age, weight, and underlying disease. Patients with cardiac dysfunction or PHTN may benefit from a narcotic- and benzodiazepine-based strategy that minimizes vasoplegia, and better maintains blood pressure.

5.6 Is Induction Slow and Titrated or Rapid Sequence?

Anesthesiologists have differing opinions regarding the ideal induction strategy for lung transplantation owing to the clear conflict between maintaining hemodynamic stability and minimizing apneic time. In most cases, a titrated induction supported by bag mask ventilation (BMV) achieves both anesthetic goals. Exceptions to this rule include patients with predictors of difficult BMV, inadequate fasting time, or aspiration risk such as esophageal dysmotility in connective tissue disease. Even in patients receiving a titrated induction, consider using rapid sequence doses of muscle relaxants to minimize time to intubation.

Table 5.1 Common induction drugs and doses for lung transplant

- Midazolam (0.01–0.05 mg/kg)
- Fentanyl (max 2-5 ug/kg)
- Sufentanil (0.04–0.01 μ/kg)
- Ketamine (0.25–0.5 mg/kg)
- Propofol (0.5-1 mg/kg)
- Rocuronium (0.6–1.2 mg/kg)
- Succinylcholine (1.5 mg/kg)
- Hydromorphone (0.01–0.03 mg/kg)

5.7 Does Ketamine Increase PVR?

Despite favorable effects on bronchomotor tone, thoracic anesthesia dogma maintains that ketamine increases pulmonary vascular resistance and should be avoided in patients with pulmonary hypertension [1–3]. Early animal studies produced conflicting results, and human studies were conducted with high doses (2 mg/kg) in spontaneously breathing volunteers, not accounting for the effects of hypoxia and hypercarbia resulting from sedation and partial airway obstruction on pulmonary vascular resistance (PVR) [4]. Subsequent studies under controlled conditions demonstrate no increase in PVR and a favorable safety profile in patients with pulmonary hypertension, although many of these studies involve pediatric populations [5–10]. Anecdotally, several thoracic anesthesiologists at TGH routinely use ketamine as part of their induction for chronic thromboembolic pulmonary hypertension (CTEPH) patients with severe pulmonary hypertension to good effect. The benefits of ketamine include analgesia and hypnosis while maintaining stable hemodynamics and perfusion pressure [11]. When used in small doses (0.25 mg/kg), and in combination with propofol or sevoflurane, there is negligible increase in PVR. However, ketamine use remains controversial, and stable hemodynamics can similarly be achieved with a combination of other induction agents and vasopressors.

5.8 Are Defibrillator Pads Used During Lung Transplant?

Place defibrillator pads prior to induction for any patient with significant coronary artery disease, cardiomyopathy, RV dysfunction, near-systemic PA pressures, or implantable cardioverter defibrillator (ICD) in situ. Pacer-dependent patients should have their device reprogrammed to an asynchronous mode prior to the start of surgery. Pads must be placed out of surgical field, yet still allow electrical current to cross the heart. A practical approach is to keep conventional pad placement throughout induction and line placement, with removal just prior to incision. Once the clamshell incision is performed, retraction of the chest wall makes defibrillation via external pads less effective, but the surgeons can then perform internal defibrillation via paddles.

5.9 Can Milrinone and Epinephrine Infusions Be Used to Support Patients with Pulmonary Hypertension Through Induction?

The use of milrinone and epinephrine infusions during induction for lung transplantation is uncommon at TGH.

Milrinone Milrinone is a phosphodiesterase 3 inhibitor and powerful inodilator. Although milrinone effectively reduces PVR, it often does so at the expense of systemic pressures. The effect on blood pressure is unpredictable and with a long duration of action. For these reasons, milrinone is rare in lung transplant unless already infusing prior to the OR.

Epinephrine If RV dysfunction is moderate to severe, low-dose epinephrine infusion or boluses can be considered for inotropic support during induction. However, epinephrine can significantly increase pulmonary pressures, compromising any intended benefit to the RV.

5.10 What Is the Induction Strategy for Patients with Severe Pulmonary Hypertension or RV Failure?

Consider awake lines and establish a clear rescue plan with surgical team. The staff surgeon and perfusionist must be in the OR during induction in case of hemodynamic collapse. Consider having two staff anesthetists if possible. Options include:

- Titrated induction with vasopressors running, favoring hypertension throughout induction. Surgical team scrubbed and patient prepped.
- Awake groin cannulation with guidewires prior to induction so that ECMO can be quickly initiated if necessary. Consider for hemodynamically stable patients with supra-systemic PA pressures.
- Fully awake ECMO cannulation and initiation of ECLS prior to induction. Consider for hemodynamically unstable patients in respiratory distress.

A failing RV may require inotropic support to maintain perfusion pressure. Recognize that inotropes such as epinephrine will also cause significant increase in PA pressures and should be used with caution (See **Question 5.9**). All precipitants of PVR must be avoided (hypoxia, hypercarbia, acidosis, pain, light anesthesia). Consent for TEE should be documented, and the TEE machine in the room on standby.

5.11 Why Must the Staff Surgeon Be Present for Induction?

The staff surgeon must be present for every induction, as they can rapidly initiate ECLS in the event of severe instability or arrest.

5.12 What Is the Management Algorithm for Arrest on Induction?

There is no adult cardiac life support (ACLS) algorithm specifically for the management of cardiac arrest during lung transplant. Management should be directed at (1) correcting the precipitant, (2) high-quality resuscitation as per ACLS protocols, and

(3) rapid institution of extracorporeal life support (ECLS) if no return of spontaneous circulation (ROSC).

In an arrest during induction, intubate with a single lumen tube and ventilate with 100% FiO2 to eliminate hypoxia or hypercarbia as potential precipitants/contributors. Avoid hyperinflation which may compromise venous return and drug circulation. If hyperinflation is suspected, briefly disconnect the ventilator to allow full expiration. The surgical team should begin chest compressions while the nursing staff apply defibrillator pads. Consider early administration of inotropes to restore coronary perfusion pressure. Perform a pulse check and identify the rhythm. If the arrest is associated with a shockable dysrhythmia, defibrillate as soon as possible. In the interim, the surgeons should prepare for emergent initiation of ECLS. ECLS may be peripheral or central. Peripheral ECLS cannulation minimizes interruption to chest compressions but takes longer compared to central ECLS. Peripheral cannulation time is minimized by elective pre-induction wiring of femoral vessels in high-risk patients. Central ECLS is achieved through emergent chest opening via clamshell. Although this necessitates interruption of chest compressions, the surgeons can continue with internal cardiac massage soon after incision. External defibrillation will be ineffective after clamshell, and internal paddles should be used.

Once the chest is open, the type of ECLS employed will depend on the surgeon and situation. If the precipitant is unclear and arrest persists, some surgeons will choose full CPB over VA-ECMO to better control hemodynamics.

Following return of spontaneous circulation (ROSC), the team must decide whether to proceed with transplant. If there is a clear precipitant, the arrest time is minimal, and hemodynamics have stabilized with modest support, it is reasonable to continue with transplant. Cancelling the transplant is a difficult, multidisciplinary decision that weighs graft waste (a backup recipient may not be readily available) against the potential for poor outcome in the current recipient, who will likely not wean from mechanical ventilation easily.

References

1. Gooding JM, et al. A physiologic analysis of cardiopulmonary responses to ketamine anesthesia in noncardiac patients. Anesth Analg. 1977;56(6):813–6.
2. Tweed WA, et al. Circulatory responses to ketamine anesthesia. Anesthesiology. 1972;37(6):613–9.
3. Spotoft H, et al. The cardiovascular effects of ketamine used for induction of anaesthesia in patients with valvular heart disease. Can Anaesth Soc J. 1979;26(6):463–7.
4. Maxwell BG, Jackson E. Role of ketamine in the management of pulmonary hypertension and right ventricular failure. J Cardiothorac Vasc Anesth. 2012;26(3):e24–5. author reply e25-6
5. Williams GD, et al. Ketamine does not increase pulmonary vascular resistance in children with pulmonary hypertension undergoing sevoflurane anesthesia and spontaneous ventilation. Anesth Analg. 2007;105(6):1578–84. table of contents
6. Williams GD, et al. Perioperative complications in children with pulmonary hypertension undergoing general anesthesia with ketamine. Paediatr Anaesth. 2010;20(1):28–37.
7. Basagan-Mogol E, et al. Induction of anesthesia in coronary artery bypass graft surgery: the hemodynamic and analgesic effects of ketamine. Clinics (Sao Paulo). 2010;65(2):133–8.

8. Oklu E, et al. Which anesthetic agent alters the hemodynamic status during pediatric catheterization? Comparison of propofol versus ketamine. J Cardiothorac Vasc Anesth. 2003;17(6):686–90.
9. Hickey PR, et al. Pulmonary and systemic hemodynamic responses to ketamine in infants with normal and elevated pulmonary vascular resistance. Anesthesiology. 1985;62(3):287–93.
10. Friesen RH, et al. Hemodynamic response to ketamine in children with pulmonary hypertension. Paediatr Anaesth. 2016;26(1):102–8.
11. Reimer C, Granton J. Pharmacology of the pulmonary circulation. In: Slinger P, editor. Principles and practice of anesthesia for thoracic surgery; 2019. p. 165–79.

Airway Management

6

Abstract

Airway management for lung transplantation must take into consideration limited reserve, rapid desaturation, and altered respiratory physiology. A thorough airway assessment is essential for predicting anatomical challenges. Video laryngoscopy may enhance glottic visualization and reduce intubation time for double-lumen endotracheal tubes. For patients with tracheostomies or those bridged on extracorporeal membrane oxygenation (ECMO), individualized approaches to lung isolation are essential to ensure procedural success.

6.1 How Are Recipients Ventilated During Induction?

Bag mask ventilation (BMV) during induction is crucially important for lung transplant recipients with marginal reserve. Aim to minimize apneic time to avoid hypoxemia and hypercapnia which can exacerbate pulmonary pressures. Note that these patients may desaturate **in seconds**, so consider routine use of oropharyngeal airways to maximize the likelihood of successful BMV on first attempt. Low lung compliance often causes high resistance to BMV, causing high airway pressure and risk of gastric insufflation and aspiration. Match your BMV respiratory pattern to the patient's underlying pathology. Patients with restrictive lung disease are best bag mask ventilated with small tidal volumes and higher rates Vt 4-6 cc/kg (~350-400 cc) x 16 bpm. Conversely, patients with obstructive lung disease are better served with normal tidal volumes and low rates, Vt 6-8 cc/kg (~400-450 cc) x 10 bpm, allowing full expiration between breaths to prevent breath stacking and hyperinflation (see **Question 5.3**). Regardless of induction strategy, use rapid sequence doses of muscle relaxants to minimize time to intubation.

M. Salvatori, A. Huang (eds.), *Anesthesia for Lung Transplant*,
https://doi.org/10.1007/978-3-032-06112-6_6

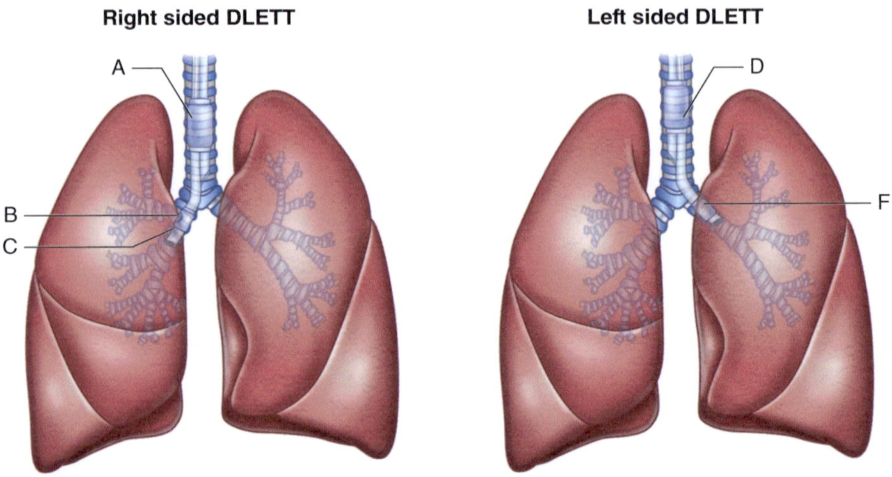

Fig. 6.1 DLETT positioning [2] **A**. right sided DLETT tracheal cuff, **B**. Murphy eye for right upper lobe ventilation, **C**. right sided DLETT bronchial cuff, **D**. left sided DLETT tracheal cuff, **E**. left sided DLETT bronchial cuff

Table 6.1 DLETT sizing [1]

Sex	Height (cm)	DLETT size (Fr)
Female	<160 cm	35
Female	≥ 160 cm	37
Male	<170 cm	39
Male	≥ 170 cm	41

6.2 How Are DLETTs Sized and Positioned?

As with most thoracic surgery, transplants are almost exclusively performed with left-sided double-lumen tubes (DLETT). Surgeons may rarely request a right-sided double-lumen tube for left single-lung transplants. Figure 6.1 depicts the optimal position for right and left double-lumen tubes. Table 6.1 shows DLETT size recommendations as per *Miller's Anesthesia* [1]. There are differing opinions among anaesthesiologists regarding upsizing vs. downsizing tubes for patients near the height cutoffs. Upsizing to a larger tube provides a tighter fit, less leak, and easier bronchoscopy and suctioning. Downsizing to a smaller tube permits easier insertion, less airway trauma, and reduced risk of left upper lobe bronchus occlusion. Disease processes may also impact airway calibre, with fibrotic lung disease (ILD, CF) associated with more retracted airways, whereas COPD airways may be expanded.

End-stage lung disease can result in significant changes in airway anatomy. Both restrictive and obstructive pathologies can cause traction and displacement resulting in tracheal deviation, airway narrowing, and/or acute angulation of the left main bronchus. Radiologists may not explicitly comment on these changes, and the CT

Table 6.2 Algorithm for difficult DLETT positioning

• Apply manual pressure to the neck to displace the trachea or DLETT as required. • Use the bronchoscope as a stylet through the bronchial lumen to guide the DLETT to the left main bronchus. This is more challenging with smaller DLETTs (35-39Fr) that only accommodate a pediatric bronch, which is extremely pliable and ineffective as a stylet. • Change to a Silbroncho DLETT—a soft-tipped, flexible, silicone DLETT that may better accommodate acute angulations of the left main bronchus. • Intubate and isolate the lungs using a single-lumen tube over an adult bronchoscope. Once in the left main bronchus, use an airway exchange catheter to intubate with a DLETT. • As a last resort, intubate with a SLETT and proceed with VA ECMO. Once the right mainstem bronchus is open, the surgeons may be able to manually guide the endobronchial lumen to the left main bronchus.

chest images should be reviewed as part of the preoperative assessment. Gross CT abnormalities predict challenging endobronchial intubation that may require advanced techniques, listed in Table 6.2.

6.3 What Are the Lung Isolation Options for Difficult Airways?

Poor respiratory reserve and reduced time to desaturation exacerbate the challenges of a known or suspected difficult airway. In these cases, the safest strategy is awake fibreoptic intubation with a single lumen tube. Awake fibreoptic DLETT intubation is described in the literature [3] but generally avoided due to patient discomfort. Once the patient is intubated, proceed with induction of anesthesia, followed by DLETT tube exchange over a soft-tipped airway exchange catheter. If the airway is distorted or edematous and there is risk of tube exchange failure, consider performing the exchange after the patient is on ECMO.

6.4 Is DLETT Intubation Easier with Direct or Video Laryngoscopy?

A 2021 systematic review and meta-analysis of 25 studies (2154 patients) compared direct laryngoscopy (DL) vs. video laryngoscopy (VL) for DLETT intubation in thoracic surgery [4]. They found that VL offers shorter intubation time (43 vs. 54 s), better glottic view, and a lower need for external laryngeal manipulation (ELP) but comparable first intubation attempt success rate and overall intubation success rate compared with DL. A 2018 systematic review and meta-analysis of 14 studies (310 patients) showed that VL resulted in higher success rate at first attempt and a lower incidence of intubation-related complications, but with comparable intubation times [5]. However, VL was also associated with a higher incidence of DLETT malposition (9.9% vs. 4.5%).

Although these studies have discrepant findings, it appears that VL is at least non-inferior to DL, and at best provides some advantages, including shorter intubation times, higher success rate (particularly for less experienced operators), lower incidence of complications, and allowing both the anesthesia and surgical teams to visually confirm DLETT placement.

6.5 What Are the Lung Isolation Options for a Tracheostomy?

In patients with otherwise normal airway anatomy, exchange the tracheostomy for an oral DLETT and isolate conventionally. If the upper airway is grossly abnormal (i.e., radiation injury) or discontinuous (i.e., laryngectomy), the tracheostomy stoma is directly intubated with a single lumen tube, and isolation achieved with a bronchial blocker.

6.6 How Is the Airway Managed For Patients Already on ECMO?

Recipients bridged to transplant with ECMO require DLETT intubation and conventional lung isolation. A tube exchange catheter is advised for patients intubated for over 5 days in MSICU. Although theoretically, the early stages of transplant can be performed on ECMO without an airway, later stages always require selective lung isolation.

References

1. Slinger P, Campos J. In: Miller R, editor. Anesthesia for thoracic surgery. Miller's Anesthesia: Elsevier; 2015.
2. Bora V, et al. Double lumen endobronchial tubes. 2023. https://www.statpearls.com/ArticleLibrary/viewarticle/20690. Accessed 19 Feb 2023.
3. Motamed C, Salazar G. Awake fiberoptic intubation with target-controlled infusion of remifentanil in emergency surgery. J Opioid Manag. 2010;6(1):11.
4. Karczewska K, et al. Efficacy and safety of video-laryngoscopy versus direct laryngoscopy for double-lumen endotracheal intubation: a systematic review and meta-analysis. J Clin Med 2021;10(23).
5. Liu TT, et al. Videolaryngoscopy vs. Macintosh laryngoscopy for double-lumen tube intubation in thoracic surgery: a systematic review and meta-analysis. Anaesthesia. 2018;73(8):997–1007.

Central Lines

7

Abstract

Standard central access for lung transplantation includes a right internal jugular Cordis with pulmonary artery catheter and a left subclavian central line. Left-sided access preserves the right IJ for possible ECMO cannulation. Ultrasound guidance for subclavian access reduces complication rates, particularly for inexperienced operators. Common complications such as arterial injury and iatrogenic pneumothorax can lead to major instability if not recognized.

7.1 What Are the Standard Lines for Lung Transplant?

The conventional line arrangement for double lung transplants is a right internal jugular (IJ) Cordis and Swan-Ganz catheter, and a left subclavian triple lumen central venous catheter (CVC). In complex cases with a higher likelihood of post-operative graft dysfunction, consider placing both the Cordis and triple lumen CVC on the left. This leaves the right IJ free for planned or emergent post-operative veno-venous (VV) ECMO cannulation. For the same reason, avoid placing both lines on the right if possible. If subclavian access proves difficult, consider bilateral IJ cannulation. Recall that the triple-lumen CVC is primarily used for post-transplant care in the ICU and transplant stepdown. Do not delay the transplant if you are unable to secure the second line and have adequate IV access otherwise. In single lung transplants, it is convention to place the subclavian line on the side to be transplanted, in case of iatrogenic vascular or lung injury.

The standard triple-lumen and Cordis catheters at TGH have an antimicrobial coating containing both chlorhexidine and silver sulfadiazine. PA catheters are not antimicrobial-coated but do contain latex. Drug or latex-free lines are available in the sterile core for patients with known allergies.

M. Salvatori, A. Huang (eds.), *Anesthesia for Lung Transplant*,
https://doi.org/10.1007/978-3-032-06112-6_7

7.2 Is There Any Benefit to Using Ultrasound for Subclavian Line Insertion?

Subclavian cannulation is often safe and uncomplicated using an anatomical landmark approach. Complication rates vary from 0.3 to 12% depending on operator experience [1]. Reported complications include failure to locate or cannulate the vein, puncture of the subclavian artery, misplacement of the catheter (defined as placement of the catheter tip in the contralateral subclavian vein or in either jugular vein), pneumothorax, mediastinal hematoma, hemothorax, and injury to adjacent nerves [2].

In the first study to compare ultrasound-guided vs. landmark-guided subclavian cannulation in 821 patients, Mansfield et al. (1994) showed that ultrasound guidance had no effect on the rate of complications or failures of subclavian-vein catheterization [1]. Predictors of failure included previous major surgery in the area, a body mass index (BMI) of >30 or < 20, and previous catheterization. Of critical importance to our practice, the complication rate rose from 4% with one needle attempt, to 24% with >2 needle attempts. A small study by Gaultieri et al. (1995) in 33 patients showed that ultrasound improved rates of successful catheterization in less experienced operators [2]. Specifically, novice operators had only a 44% success rate with a landmark technique compared to 92% in the ultrasound group. Ultrasound salvage of failed landmark attempts was successful in 80% of cases. Lefrant et al. (1998) compared subclavian cannulation using pulsed Doppler ultrasound vs. landmark technique in 286 patients and showed fewer misplaced catheters and a reduced rate of mechanical complications using ultrasound, although time taken was significantly longer [3]. Fragou et al. [4] reviewed line placement in 400 mechanically ventilated patients and showed that ultrasound-guided subclavian cannulation was successful in 100% of cases, compared with 87.5% in the landmark group [4]. Ultrasound guidance was also associated with fewer attempts and fewer mechanical complications.

In summary, these studies suggest that ultrasound guidance may increase success rate and reduce complications during subclavian cannulation, especially for novice operators. At a minimum, we recommend becoming familiar with ultrasound-guided subclavian cannulation for rescue after a failed landmark approach. Given the results of the Mansfield study, no more than 2 attempts should be made by a single operator using the landmark technique.

7.3 How Is Iatrogenic Subclavian Artery Injury Managed?

The signs of inadvertent arterial puncture are brisk backflow of bright-red, oxygenated blood into the syringe upon aspiration, and pulsatile flow from the needle when the syringe is removed. In lung transplant recipients with poor ventilatory capacity and low PaO2, arterial blood may appear dark, resembling venous blood. Likewise, in hypotensive patients, the pulsatility of the backflow may not be immediately apparent. Consider manometry confirmation for all subclavian line insertions, but especially for patients in which access was difficult. When confirming via manometry, use a longer 18-gauge angiocath to prevent inadvertent decannulation. Ezaru et al. (2009) showed that manometry during CVC cannulation completely eliminated arterial dilatation in a study of 9348 patients over 15 years [5]. Alternatives to manometry include blood gas analysis or transducing prior to dilatation.

If cannulation of the subclavian artery is detected following dilatation and CVC insertion, leave the central line in situ and immediately consult vascular surgery. In most cases, the subclavian artery will be amenable to repair from inside the chest following clamshell incision and pneumonectomy. Be prepared to transfuse to replace ongoing losses.

7.4 How Is Iatrogenic Lung Injury Managed?

Iatrogenic lung injury during subclavian cannulation will not affect the long-term outcome of a double lung transplant but may cause emergent instability prior to chest opening due to tension pneumothorax or ventilatory compromise in patients with limited reserve. The first indicator of iatrogenic lung injury is often a bloodless loss of vacuum in the syringe. Sudden changes in ventilatory parameters such as peripheral oxygen saturation (SpO2), oxygen requirements, airway pressures, and maximal achievable tidal volume are also early indicators of pneumothorax. At this stage, lung ultrasound may show a lung point or absence of lung sliding suggestive of an evolving pneumothorax. Notably, lung ultrasound is superior to supine chest x-ray for the detection of pneumothorax, with sensitivity of 90.9% and specificity of 98.2% (vs. 50.2% and 99.4% respectively for supine CXR) [6].

A late but ominous sign is unexpected hypotension following line insertion. At this stage, clinical exam findings such as unilateral chest expansion and breath sounds should be evident. Temporize the situation with volume and pressors while the surgeons relieve the intrathoracic pressure, either via expedited clamshell incision or needle decompression and chest tube insertion.

7.5 Should Old Central Lines In Situ Be Removed?

By convention, central lines in place for >72 hours should be replaced at the time of transplantation to reduce the risk of bacteremia. However, this recommendation is not evidence-based, as it has long been recognized that routine central line exchanges do not reduce the risk of central line-associated bloodstream infections (CLABSI) [7]. As summarized by Timsit (2000): "The daily risk of CVC infection is considered to be a constant; the risk of catheter infection is directly related to the duration of catheter insertion. Consequently, the routine change of the catheter is able to decrease the number of infections per catheter but not to modify the number of infections per day of catheter insertion" [8].

7.6 How Do I Insert and Position a PA Catheter (PAC)?

PAC insertion is through the hemostasis port of a Cordis catheter. Never float a PAC through an old Cordis line, except in extreme emergencies. Following Cordis insertion, doff the outer set of sterile gloves and place a new mini-drape over the soiled field to prevent PAC contamination. Pass the PAC through the sterile sheath in the correct orientation to ensure that the distal end locks into the cordis. With the help of your assistant, confirm the patency of all 3 pressure/injection ports and assess balloon integrity. Although insertion can occur from any venous cannulation site, the RIJ vein provides the most direct route to the right heart chambers. Insert the PAC in the Cordis, advance to 20 cm, inflate, and clamp the balloon. The balloon allows flow-directed advancement of the PAC into the RA, through the tricuspid valve (TV) into the RV and through the pulmonic valve into the main PA. Advance the PAC using the transduced trace from the distal port as a guide (Fig. 7.1). The air-filled balloon floats to nondependent regions as it passes through the heart. The head down position aids flotation past the TV and tilting to the right with head up encourages passage out of the RV. From a right internal jugular (RIJ) vein puncture site, the PAC reaches the RA at 20–25 cm, RV at 30–35 cm, PA at 40–45 cm, and wedge at 45–55 cm. Add an additional 5–10 cm when using the left IJ vein. The PA trace is an arterial waveform that closely mirrors the systemic arterial pressure waveform. Pulmonic valve closure causes the dicrotic notch. Normal PA systolic pressures are 15-25 mmHg, PA diastolic pressures 8-15 mmHg and mean PAP

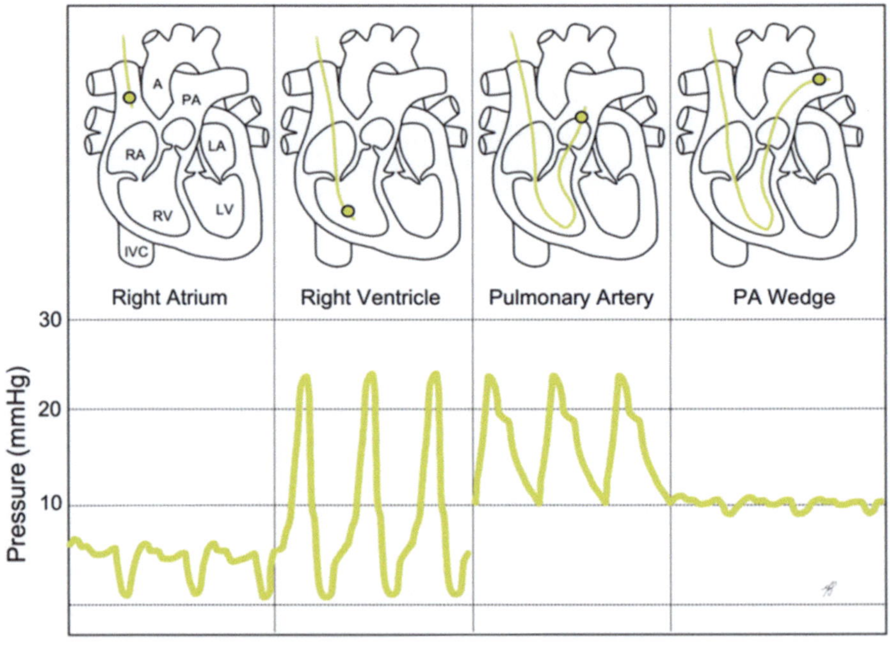

Fig. 7.1 Pressure tracing during PAC insertion

10-22 mmHg. Pulmonary hypertension (PH) by definition, is a mean PAP > 20 mmHg (sixth World Symposium on PH) [9].

If the expected trace does not appear after reaching these average distances, suspect coiling or misdirection of the PAC. Deflate the balloon, withdraw the catheter to 20 cm, and repeat the PAC floating sequence. In cases where catheter advancement proves difficult, TEE can guide the operator. Following correct catheter positioning in the main PA 1-2 cm distal to the pulmonic valve, deflate the balloon, and lock the catheter in place at the Cordis hub.

7.7 What Steps Can Be Taken to Troubleshoot a PA Catheter? (Table 7.1)

Table 7.1 Troubleshooting PAC Insertion

Issue	Possible causes	Corrective steps
Unable to enter RV *(CVP trace persists past 30 cm)*	• PA pressure cable is connected to CVP • Coiling in the RA • Chiari network • Tricuspid stenosis or regurgitation • Advancing past RA into IVC	• Verify that the pressure cable is attached to the correct transducer • Re-zero and set correct scale on monitor • Place the patient head down • Consider TEE guidance
Unable to enter PA *(RV trace persists past 50 cm)*	• Coiling in the RV • Severe PR	• Place the patient in head up and tilt right. This makes the RVOT the least dependent area of the RV • Rotate the catheter while advancing • Transiently increase BP with pressors (SBP 100-120 mmHg) • Consider TEE guidance
Unstable arrythmias	• Irritation of the RV endocardium	• Ensure the defibrillator pads are connected and capturing • Deflate the balloon and withdraw to the right atrium • If arrythmia persists, correct any electrolyte abnormalities. Consider magnesium, lidocaine, and/or amiodarone • Defer insertion until the chest is open in case emergency cannulation is required
CVP trace reappears after successful insertion	• Inadvertent withdrawal back to the RA (line not locked) • Inadvertent advancement into distal PA (wedge)	• Confirm that the PAC depth is unchanged • Ensure the balloon is deflated and withdraw 1–2 cm • Lock the PAC in place

References

1. Mansfield PF, et al. Complications and failures of subclavian-vein catheterization. N Engl J Med. 1994;331(26):1735–8.
2. Gualtieri E, et al. Subclavian venous catheterization: greater success rate for less experienced operators using ultrasound guidance. Crit Care Med. 1995;23(4):692–7.
3. Lefrant JY, et al. Pulsed Doppler ultrasonography guidance for catheterization of the subclavian vein: a randomized study. Anesthesiology. 1998;88(5):1195–201.
4. Fragou M, et al. Real-time ultrasound-guided subclavian vein cannulation versus the landmark method in critical care patients: a prospective randomized study. Crit Care Med. 2011;39(7):1607–12.
5. Ezaru CS, et al. Eliminating arterial injury during central venous catheterization using manometry. Anesth Analg. 2009;109(1):130–4.
6. Alrajhi K, et al. Test characteristics of ultrasonography for the detection of pneumothorax: a systematic review and meta-analysis. Chest. 2012;141(3):703–8.
7. Cobb DK, et al. A controlled trial of scheduled replacement of central venous and pulmonary-artery catheters. N Engl J Med. 1992;327(15):1062–8.
8. Timsit JF. Scheduled replacement of central venous catheters is not necessary. Infect Control Hosp Epidemiol. 2000;21(6):371–4.
9. Galiè N, McLaughlin VV, Rubin LJ, et al. An overview of the 6th World Symposium on Pulmonary Hypertension. Eur Respir J. 2019;53:1802148. https://doi.org/10.1183/13993003.02148-2018.

Anesthesia and Analgesia

8

Abstract

Anesthesia for lung transplantation may be maintained using volatile agents or total intravenous anesthesia (TIVA). Volatile anesthetics may confer protection against ischemia-reperfusion injury, but should not be used as sole agents due to inconsistent uptake during ventilation interruptions and impaired perfusion. TIVA with propofol ensures consistent delivery and is preferred during ECMO. Opioids, typically hydromorphone or sufentanil, are used variably; sufentanil infusions offer hemodynamic stability with a favorable pharmacokinetic profile. Analgesia adjuncts like ketamine or lidocaine may benefit select patients. Thoracic epidurals improve outcomes but carry bleeding risks, while plane blocks are promising alternatives. Pre- and post-transplant opioid use impacts recovery and survival.

8.1 How Is Anesthesia Maintained During Lung Transplant?

Anesthesia during transplant can be safely and reliably maintained with either volatile anesthetics or total intravenous anesthesia (TIVA). All volatile anesthetics impair hypoxic pulmonary vasoconstriction (HPV) in a dose-dependent manner, although this effect is only clinically significant with older agents such as halothane [1]. In animal studies, sevoflurane does not impair HPV at a concentration of 1 Minimum Alveolar Concentration (MAC) [2]. Sevoflurane, desflurane, and isoflurane all impair HPV to a similar extent [3] Volatile anesthetics are safe to use and will not negatively affect oxygenation during OLV [4].

There is some evidence that volatile anesthetics effectively suppress local alveolar inflammatory responses to OLV and thoracic surgery [5, 6]. Research conducted at the TLTP showed that either preconditioning or postconditioning of lungs with 1.5% sevoflurane improved the oxygenation of lung grafts, protecting against

M. Salvatori, A. Huang (eds.), *Anesthesia for Lung Transplant*,
https://doi.org/10.1007/978-3-032-06112-6_8

ischemia-reperfusion injury in a rat model [7]. These findings have yet to be replicated in human studies or clinical trials.

Volatiles should not be used alone for maintenance of anesthesia during lung transplantation due to inadequate and inconsistent delivery and uptake. During OLV with the patient's native, diseased lung, uptake may be poor and expired concentrations may not reflect blood concentrations in the context of significant V/Q mismatch [8]. However, COVID research demonstrated that effective blood concentrations of inhaled anesthetics can be achieved even with severe acute respiratory distress syndrome (ARDS), provided the necessary concentration gradient is maintained [8]. More concerning is that ventilation is frequently interrupted throughout the transplant for bronchoscopy and suctioning. Tears and leaks in the diseased lung are not uncommon during dissection and pneumonectomy, and volatile contamination of the OR may occur. For lung transplants on ECMO, lung perfusion is inconsistent and often impaired by hypotension or PA compression, which compromises uptake of volatiles.

Alternatively, TIVA with propofol +/− opioids can provide reliable anesthesia regardless of interruptions in ventilation. Administer TIVA through central lines to minimize the risk of occult extravasation in peripheral veins. Do not use the distal PAC port for delivery of propofol or other drugs, as the tip of the PAC may be removed from the pulmonary artery during the PA anastomosis.

Incorporating all these considerations, a conservative approach would be to administer sevoflurane of at least 1.5% throughout the transplant for protection against ischemia-reperfusion injury, with a background of low-dose propofol (25-50mcg/kg/min) to cover any short interruptions in volatile delivery. Monitor the processed EEG throughout the transplant and titrate anesthetics accordingly.

8.2 Are There Any Significant Differences Between Volatiles?

Isoflurane has greater blood solubility than sevoflurane or desflurane [9]. Some authors have theorized that the pharmacokinetics of soluble isoflurane will be less affected by severe lung disease and V/Q mismatch [8]. However, during OLV, no significant differences were found between 1 MAC isoflurane and 1 MAC sevoflurane with respect to PaO_2, shunt, cardiac output, or other hemodynamics [10]. Furthermore, only sevoflurane has been shown to protect against lung ischemia-reperfusion injury in animal models [7].

8.3 What Is the Average Total Narcotic Dose?

Practice varies among clinicians in the lung transplant anesthesia group. Opioid-naïve patients typically receive 1-2 mg of Hydromorphone IV. There is significant variation in timing, but administration typically occurs toward the later stages of the transplant.

Currently, there are no studies on intraoperative opioid practices in lung transplant recipients. Most lung transplant recipients remain sedated, intubated, and ventilated for >12 h. As such, intraoperative opioids likely have a limited effect on the outcome and quality of recovery. Furthermore, the ICU protocol at TGH is to initiate a fentanyl infusion shortly after transfer, as part of their analgesia and sedation protocol.

8.4 Are There Any Benefits to Sufentanil Infusions?

There is no proven benefit to opioid infusions over bolus dosing during lung transplant, and the choice depends on clinician preference. In a global survey of 75 lung transplant centers, sufentanil was used for intraoperative analgesia in 42.7% of centers, more commonly in Europe [11].

Sufentanil infusions (typical dose range 20-30mcg/hr) provide a steady plasma concentration of potent opioid which may minimize hemodynamic reactivity to surgical stimulus. Infusions also allow dose titration to match the changes in stimulation intensity. Avoid prolonged, high-dose infusions, as sufentanil-induced hyperalgesia has been reported [12]. Sufentanil has no active metabolites of concern.

Sufentanil has a predictable context-sensitive half-life of approximately 1-hour for infusions of 6–10 h in duration [13]. When compared to fentanyl, sufentanil is associated with less respiratory depression despite equivalent analgesia [14]. Sufentanil's pharmacokinetic profile is advantageous in lung transplant, as infusions discontinued at the end of surgery provide analgesia for transfer without impacting sedation or ventilation in the ICU.

8.5 Are Ketamine or Lidocaine Effective Pain Adjuncts?

Few studies have investigated the use of ketamine and lidocaine for lung transplant or clamshell incisions. Multimodal analgesia protocols involving both ketamine and lidocaine for recipients ineligible for an epidural have been described [15]. Although low-dose ketamine has been shown to reduce opioid consumption and frequency of nocturnal desaturations after thoracotomy [16], the benefit of ketamine in reducing acute and chronic pain after thoracic surgery remains controversial, with little evidence supporting routine use [17]. A single case report describes IV lidocaine infusion for intractable incisional pain following a lung transplant for cystic fibrosis [18]. Although the lidocaine infusion improved analgesia and reduced opioid consumption, no other non-opioid adjuncts or regional techniques were considered. Overall, the evidence does not support ketamine or lidocaine as first-line agents in lung transplant. However, these adjuncts may benefit patients with refractory pain, excessive opioid requirements, or opioid-related side effects.

8.6 Is IV-PCA Routine After Lung Transplant?

At the TLTP, post-transplant IV-PCA is common, typically starting on postop day 2–4 after extubation. In select cases, IV-PCA can be initiated as the patient approaches extubation readiness, in coordination with the ICU and transplant teams. Patients who remain sedated and intubated on postoperative days 3–4 will typically not require IV-PCA after a prolonged vent wean. Due to concerns regarding opioid-related sedation and hypoventilation, IV-PCA orders are only initiated by the APS team after a thorough assessment. IV-PCA is not pre-emptively ordered during the OR.

8.7 Is there a Role for Neuraxial or Regional Analgesia?

Thoracic epidurals in lung transplant have been extensively studied regarding both analgesia and transplant outcomes. Several large studies and structured reviews have demonstrated that thoracic epidurals improve analgesia, reduce ventilation time, and shorten ICU length-of-stay [19–24]. These benefits are comparable with either pre- or postoperative placement.

Despite these demonstrable benefits, neuraxial techniques are underutilized due to the associated risks. Transplants are increasingly being performed on ECMO, necessitating systemic anticoagulation which increases the risk of epidural hematoma. Similarly, post-transplant immunosuppression increases the risk of epidural infection and abscess. The conceived risks are likely exaggerated, as the largest study of epidural use in lung transplant recipients (119 patients) reported no instances of hematoma or infection [21]. In this study, CPB was commonly used, and epidurals were placed in the early postoperative period. Other common but less significant risks include hypotension, dural puncture, incomplete analgesia, pruritus, and nerve injury.

Several regional techniques have been developed as alternatives to epidurals, including the serratus anterior plane block (SAPB) and erector spinae plane block (ESPB). Consider these alternatives for patients with contradictions to neuraxial analgesia. These techniques can be used with indwelling catheters to provide continuous analgesia while avoiding the major risks of neuraxial analgesia. Evidence for SAPBs and ESPBs in lung transplant recipients is limited to case reports [25–30], all of which demonstrate an excellent safety profile, and analgesic benefit over systemic opioids. The TGH Thoracic Anesthesia Group has published promising results with ESPBs after lung transplant [28]. The impact of these blocks on extubation time and ICU length of stay have yet to be determined.

Overall, it appears safe and advantageous to employ early postoperative thoracic epidurals and adjunctive regional techniques for patients with inadequate pain control after lung transplantation.

8.8 Does Preoperative Opioid Dependence Affect Outcome?

Chronic opioid use affects lung transplant candidacy. Specifically, transplant programs may not list opioid-dependent patients over concerns regarding challenging postoperative pain management, worsening opioid tolerance, and risk of respiratory depression.

The TLTP requires potential recipients with a history of opioid use to be referred to the transitional pain service (TPS) for evaluation. Should they meet the listing criteria, the TPS facilitates weaning to a total opioid requirement of less than 100 mg oral morphine equivalents per day.

To date, no studies have investigated the impact of pre-transplant opioid use on immediate post-transplant outcomes, such as time to extubation or hospital length of stay. Vahidy et al. (2020) examined the impact of pre-transplant opioid use on retransplant free survival [31]. In this study of 425 recipients, 14% had pre-transplant opioid requirements, with a median daily morphine equivalent dose of 31 mg. Opioid use at the time of listing did not affect time to death or retransplantation but was associated with a slightly longer length of hospitalization (34 vs. 27 days).

In summary, opioid dependence poses perioperative challenges but should not preclude transplantation. Additional studies are needed to better understand and stratify the risks associated with opioid use in these patients.

8.9 What Is the Impact of Post-transplant Opioid Use?

The implications of postoperative opioid use in lung transplant recipients are an emerging area of interest that lacks consensus. In a study of 16 cystic fibrosis recipients, Drees et al. [32]. reported that 3 or more consecutive months of opioid use after transplant was associated with decreased lung function (FEV1) and increased mortality [32]. Seventy-five percent of patients in this sample were opioid-naïve prior to transplant, and the dose of opioids in chronic users was not reported.

A larger national US registry study of 11,568 recipients also found that post-transplant opioid use (but not pre-transplant) was associated with increased mortality [33]. Pre- and post-transplant benzodiazepine use was also associated with increased post-transplant mortality.

References

1. Tarry D, Powell M. Hypoxic pulmonary vasoconstriction. BJA Educ. 2017;17(6):208–13.
2. Kerbaul F, et al. Effects of sevoflurane on hypoxic pulmonary vasoconstriction in anaesthetized piglets. Br J Anaesth. 2000;85(3):440–5.

3. Pagel PS, et al. Desflurane and isoflurane produce similar alterations in systemic and pulmonary hemodynamics and arterial oxygenation in patients undergoing one-lung ventilation during thoracotomy. Anesth Analg. 1998;87(4):800–7.
4. Beck DH, et al. Effects of sevoflurane and propofol on pulmonary shunt fraction during one-lung ventilation for thoracic surgery. Br J Anaesth. 2001;86(1):38–43.
5. De Conno E, et al. Anesthetic-induced improvement of the inflammatory response to one-lung ventilation. Anesthesiology. 2009;110(6):1316–26.
6. Schilling T, et al. Effects of volatile and intravenous anesthesia on the alveolar and systemic inflammatory response in thoracic surgical patients. Anesthesiology. 2011;115(1):65–74.
7. Ohsumi A, et al. Sevoflurane attenuates ischemia-reperfusion injury in a rat lung transplantation model. Ann Thorac Surg. 2017;103(5):1578–86.
8. Kretzschmar M, et al. Effect of bronchoconstriction-induced ventilation-perfusion mismatch on uptake and elimination of isoflurane and Desflurane. Anesthesiology. 2017;127(5):800–12.
9. Esper T, et al. Blood/gas partition coefficients for isoflurane, sevoflurane, and desflurane in a clinically relevant patient population. Anesth Analg. 2015;120(1):45–50.
10. Slinger P, Scott WA. Arterial oxygenation during one-lung ventilation. A comparison of enflurane and isoflurane. Anesthesiology. 1995;82(4):940–6.
11. Tomasi R, et al. Intraoperative anesthetic Management of Lung Transplantation: center-specific practices and geographic and centers size differences. J Cardiothorac Vasc Anesth. 2018;32(1):62–9.
12. Fechner J, et al. The impact of intra-operative sufentanil dosing on post-operative pain, hyperalgesia and morphine consumption after cardiac surgery. Eur J Pain. 2013;17(4):562–70.
13. Hughes MA, et al. Context-sensitive half-time in multicompartment pharmacokinetic models for intravenous anesthetic drugs. Anesthesiology. 1992;76(3):334–41.
14. Bailey PL, et al. Differences in magnitude and duration of opioid-induced respiratory depression and analgesia with fentanyl and sufentanil. Anesth Analg. 1990;70(1):8–15.
15. Fessler J. The anesthesia Foch lung transplant protocol Suresnes. France: Foch Hospital; 2020. Available from: http://anesthesie-foch.org/wp-content/uploads/2016/02/Anesthesia-Foch-lung-transplantation-protocol.pdf2022.
16. Michelet P, et al. Adding ketamine to morphine for patient-controlled analgesia after thoracic surgery: influence on morphine consumption, respiratory function, and nocturnal desaturation. Br J Anaesth. 2007;99(3):396–403.
17. Moyse DW, et al. Perioperative ketamine Administration for Thoracotomy Pain. Pain Physician. 2017;20(3):173–84.
18. Moyano J, et al. Use of systemic lidocaine for postoperative acute pain management in single-lung transplantation: case report. Revista de la Facultad de Med 2020;68(2).
19. Verzelloni A, Walker C. An audit of postoperative use of thoracic epidural analgesia (TEA) after lung transplantation (LT). J Cardiothorac Vasc Anesth. 2018;32:S98.
20. Pottecher J, et al. Does thoracic epidural analgesia improve outcome after lung transplantation? Interact Cardiovasc Thorac Surg. 2011;12(1):51–3.
21. Cason M, et al. The efficacy and safety of epidural-based analgesia in a case series of patients undergoing lung transplantation. J Cardiothorac Vasc Anesth. 2015;29(1):126–32.
22. Bowles C, et al. Thoracic epidural analgesia improves outcomes after lung transplant. J Heart Lung Transplant. 2021;40(4):S355.
23. McLean SR, et al. Assessing the benefits of preoperative thoracic epidural placement for lung transplantation. J Cardiothorac Vasc Anesth. 2018;32(6):2654–61.
24. Axtell AL, et al. Pain management and safety profiles after preoperative vs postoperative thoracic epidural insertion for bilateral lung transplantation. Clin Transpl. 2018;32(12):e13445.
25. Kelava M, et al. Continuous erector Spinae block for postoperative analgesia after thoracotomy in a lung transplant recipient. J Cardiothorac Vasc Anesth. 2018;32(5):e9–e11.
26. Mudarth M, et al. Continuous erector Spinae plane block for analgesia after thoracotomy for lung transplantation in an anticoagulated patient. Case Rep Anesthesiol. 2021;2021:6664712.
27. Moses B, et al. Erector Spinae plane (Esp) continuous catheter for primary analgesia following bilateral lung transplant. Chest. 2020;157(6):A449.

28. Hanley C, et al. ESRA19-0562 bilateral continuous erector spinae planeblocks for the management of post-thoracosternotomy pain in patients undergoing lung transplantation—initial experience at Toronto general hospital. Reg Anesthesia Pain Med. 2019;44:A242–A43.
29. Sekandarzad MW, et al. Bilateral continuous serratus anterior blockade for postoperative analgesia after bilateral sequential lung transplantation. J Cardiothorac Vasc Anesth. 2019;33(5):1356–9.
30. Elsharkawy H, et al. Rhomboid intercostal and sub-serratus (RISS) plane block for analgesia after lung transplant. J Clin Anesth. 2019;56:85–7.
31. Vahidy S, et al. Pretransplant opioid use and survival after lung transplantation. Transplantation. 2020;104(8):1720–5.
32. Drees D, et al. Chronic opioid use and clinical outcomes in lung transplant recipients: a single-center cohort study. Clin Respir J. 2018;12(9):2446–53.
33. Lentine KL, et al. Survival implications of prescription opioid and benzodiazepine use in lung transplant recipients: analysis of linked transplant registry and pharmacy fill records. J Heart Lung Transplant. 2021;40(6):513–24.

Transplant Sequence

9

Abstract

The surgical sequence of lung transplantation differs by extracorporeal support status. For patients not on ECMO, the procedure begins with dissection and pneumonectomy of the less functional lung, followed by sequential graft implantation with bronchial, pulmonary artery, and atrial anastomoses, bronchoscopy, gradual reperfusion, and differential ventilation. The contralateral lung is then transplanted similarly. In patients on ECMO, a clamshell incision is followed by systemic heparinization, central cannulation, and initiation of extracorporeal life support. Lungs are implanted sequentially while on ECMO with careful reperfusion and ventilation. Decannulation follows, with protamine reversal as needed.

M. Salvatori, A. Huang (eds.), *Anesthesia for Lung Transplant*,
https://doi.org/10.1007/978-3-032-06112-6_9

9.1 What Is the Sequence of Events for Lung Transplants off ECMO? (Table 9.1)

Table 9.1 Lung transplant sequence off ECMO

• Surgical dissection of least functional lung
• PA test clamp with hemodynamic assessment followed by pneumonectomy
• Implantation of first graft—Anastomosis of bronchus > pulmonary artery > pulmonary veins (left atrial cuff)
• Bronchoscopy for bronchial anastomosis assessment and airway suctioning
• Reperfusion via slow release of PA clamp over 10–15 min to minimize risk of allograft injury and hypotension
• Differential lung ventilation
• Trial of OLV of the newly implanted lung
• Surgical dissection of second lung, with PA test clamp followed by pneumonectomy
• Implantation of second graft—Anastomosis of bronchus > PA > pulmonary veins
• Bronchoscopy for bronchial anastomosis assessment and airway suctioning
• Reperfusion via slow release of PA clamp over 10–15 min to minimize risk of allograft injury and hypotension
• Two lung ventilation with lung protective settings
• Hemostasis and chest closure

9.2 What Is the Sequence of Events for Lung Transplants on ECMO? (Fig. 9.1 and Table 9.2)

Table 9.2 Lung transplant sequence on ECMO

• Clamshell and opening of pericardium
• Administration of low-dose unfractionated heparin to target activated clotting time (ACT) >220 s
• Arterial cannulation of ascending aorta
• Venous cannulation of right atrium
• Initiation of ECLS, maintain ACT at 160–180 s throughout ECLS
• Sequential lung transplants (anastomosis of bronchus > pulmonary artery > pulmonary veins); bronchoscopy for bronchial assessment and suction
• Sequential reperfusion and ventilation of new lungs while still on ECLS; careful de-airing of allograft at reperfusion.
• Two lung ventilation with lung protective settings
• Consider DLT to SLT tube exchange on ECMO
• Discontinuation of ECLS and decannulation
• +/– administration of protamine depending on hemostasis and ACT
• Hemostasis and chest closure

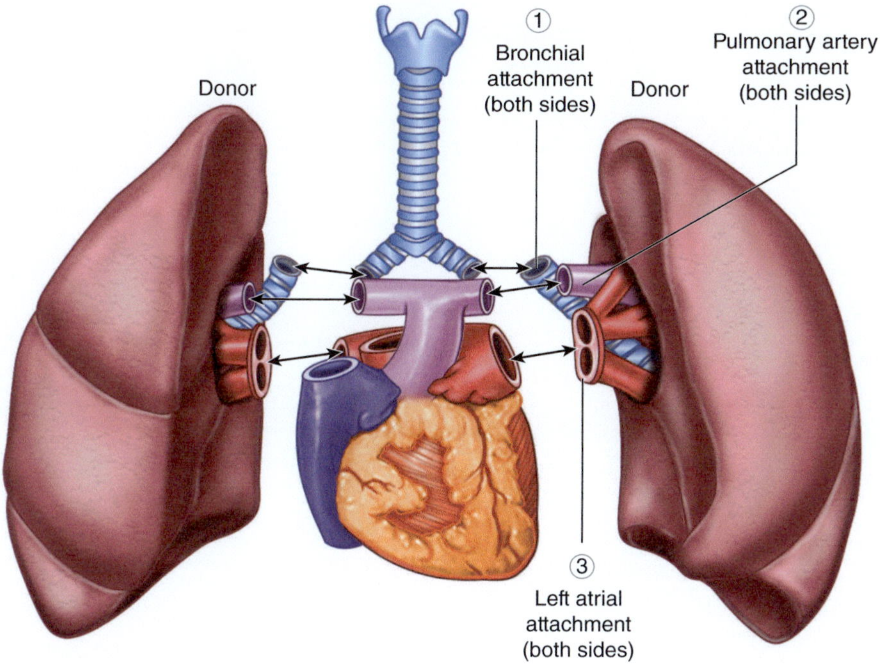

Fig. 9.1 Lung transplant sequence diagram [1]

Reference

1. Staff MC. Lung transplant. 2022. Available from: https://www.mayoclinic.org/tests-procedures/lung-transplant/about/pac-203847542023.

Ventilation 10

Abstract

Intraoperative ventilation strategies must be based on recipient physiology, with different strategies for obstructive vs restrictive disease. One-lung ventilation begins at pneumonectomy; ABG monitoring is critical due to unreliable $EtCO_2$. Air leaks, hypoxia, and poor compliance are common, with ECMO increasingly used to optimize oxygenation and hemodynamics. Inhaled nitric oxide (iNO) is reserved for severe graft dysfunction or pulmonary hypertension, though long-term benefits are uncertain. Careful coordination between anesthesia, perfusion, and surgery—including lung recruitment, bronchoscopy, and echocardiography—is essential to identify complications and guide management.

10.1 What Are the Typical Vent Settings After Induction?

The edicts of lung protective ventilation do not apply to the native lungs of the recipient, and adequate oxygenation and ventilation should be prioritized. Maintain ventilation with 100% FiO2 following intubation, which will facilitate lung collapse after isolation. Patients with COPD benefit from low ventilatory frequency (8–12 bpm), standard tidal volumes (6–8 cc/kg), and prolonged expiratory time (I:E ratio of at least 1:3, often 1:4 or longer). Be cognizant of the risk of gas trapping if EtCO2 does not return to baseline between breaths. Patients with restrictive lung disease require smaller tidal volumes (4-6 cc/kg) and higher rates (16–20 bpm) which will minimize hemodynamic compromise from high airway pressures. Do not be falsely reassured by normal EtCO2 values, as many recipients will have marked CO2 gradients due to their underlying lung disease, which worsens with positive pressure ventilation. Perform an ABG within 30 min of intubation and titrate accordingly.

M. Salvatori, A. Huang (eds.), *Anesthesia for Lung Transplant*,
https://doi.org/10.1007/978-3-032-06112-6_10

10.2 When Should One Lung Ventilation Start?

Most lung recipients will have limited ventilatory reserve, and one lung ventilation should be initiated only once the chest is open and the surgeons begin the pneumonectomy. The surgeons will typically start on the side with less lung perfusion as indicated by the preoperative perfusion scan.

10.3 What Is the Differential for Loss of EtCO2 After Clamshell? (Table 10.1)

Table 10.1 Causes of EtCO2 loss after clamshell

Ventilator and tube issues
• Cuff tear or leak
• Circuit leak, disconnect, or obstruction
• Sample line leak, disconnect, or obstruction
• Extrusion of the bronchial cuff, causing partial or complete airway obstruction
• Distal migration of DLETT
• Inadequate vent settings (i.e., pressure control ventilation resulting in inadequate tidal volumes)
Patient issues
• Thick secretions/mucous plugging
• Severe bronchoconstriction
• Low cardiac output/hypoperfusion to lungs
Surgical issues
• Tears in the native lung with loss of inspired volume
• Surgical compression of the lung, limiting expansion
• Surgical compression of the main PA, causing lung hypoperfusion
ECMO issues
• High sweep (after initiation of ECMO)
• High ECMO flows (inadequate blood flow through pulmonary circulation)

10.4 What Is the Management Algorithm for Persistent Air Leak? (Table 10.2)

Tears to the ventilated lung during dissection and pneumonectomy are not uncommon, especially in the presence of significant fibrosis or adhesions. Oxygenation will often be maintained, but the adequacy of ventilation will be difficult to assess due to poor CO2 return. Often, EtCO2 will be unreadable necessitating frequent ABG sampling. Additionally, the presence of an air leak in the setting of higher FiO_2 levels poses a risk for fire in the surgical field. The surgical team should be informed. Persistent air leak post-implantation is concerning and should prompt discussion with the surgical team and repeat bronchoscopy to assess the bronchial anastomosis.

Table 10.2 Management algorithm for persistent air leaks

- Air leak is a diagnosis of exclusion. Rule out all other immediately life-threatening differentials listed in **question 10.3**.
- Discuss your suspicions with the surgical team. They may be able to confirm the diagnosis and verify continued expansion of the ventilated lung.
- Increase FiO2 to 100% and flows to 10 L.
- Immediately perform an ABG to assess adequacy of oxygenation and ventilation. Repeat ABG q10min and titrate ventilation accordingly.
- Convert the anesthetic to full TIVA and discontinue volatiles.
- If oxygenation or ventilation are grossly inadequate, consider ECMO.

10.5 When Is Conversion to ECMO Needed for Ventilatory Reasons?

The use of intraoperative ECMO has climbed steadily at TGH, from 38% in 2012 to 61% in 2017 [1]. This trend will likely continue, as recent studies have shown that intraoperative ECMO improves 1-, 3-, and 5-year survival following bilateral lung transplant [2]. The decision to implement ECMO typically occurs in advance, especially for patients with high baseline oxygen requirements, resting hypercapnia, or pulmonary hypertension who may not tolerate one lung ventilation or clamping. However, unanticipated conversion to ECMO may be required for hypoxic respiratory failure or hemodynamic instability. Recall that it is common to experience temporary desaturation following initiation of OLV. The desaturation can be profound (SpO2 ~ 85%) and may persist until the PA is clamped, eliminating the shunt. This temporary hypoxemia does not warrant ECMO if all other parameters are stable.

Following clamping of the PA, consider conversion to VA-ECMO if you have difficulty maintaining SpO2 above 92% on 100% FiO2 or EtCO2 < 60 mmHg despite adequate tidal volumes, appropriate PEEP, and regular recruitment. These allowances may be even tighter in the context of high PA pressures. ECMO may also help in smaller patients in which surgical manipulation regularly affects hemodynamics. Express your concerns with the surgical team early to prevent the need for emergency conversion.

10.6 How Are OLV and ABG Abnormalities Managed on ECMO?

The main respiratory advantage of ECMO is that it allows control of oxygenation and partial pressure of carbon dioxide independent of ventilation. Almost all lung transplants requiring ECLS are performed using central VA-ECMO. Intraoperative VV-ECMO is only used for patients bridged preoperatively with VV-ECMO in ICU. Ventilation requirements during VA-ECMO depend on whether "full flow" can be achieved. "Full flow" indicates that the flow rate through the ECMO circuit matches native cardiac output. If full-flow ECMO is initiated from the start of surgery, ventilation settings of the native lung during pneumonectomy are immaterial. By convention, ventilation usually continues at RR 8-12 bpm, PCV 10 (Vt 4-6 cc/kg), FiO2 20–30%, and PEEP 5–10. Keep the ventilation settings constant and allow the perfusionist to manipulate ECMO settings to control PaO2 and PaCO2. If full flow cannot be achieved due to cannula size, positioning, hypovolemia, or hypotension, then a significant proportion of the blood flow will pass through the native lung, and ventilatory support may be required. Work closely with the perfusionist to establish a pH, PaO2 and PaCO2 balance.

After implantation of the first lung, ensure that you are ventilating with standard lung transplant settings: PCV 10, RR 10, and PEEP 5 on room air (FiO2 21%). Following reperfusion of the first lung, the surgeons will request a reduction in ECMO flows to allow flow through the PA. PA waveform pulsatility is a good marker of lung perfusion in the newly implanted graft.

For a more in-depth review of ECLS, see the icuECMO quick reference guide: http://icuecmo.ca/index.htm.

10.7 What Is the Checklist Prior to Reperfusion of a Newly Transplanted Lung? (Table 10.3)

Table 10.3 Pre-reperfusion checklist

- Perform a baseline ABG to confirm adequate ventilatory reserve.
- Correct any metabolic derangements, transfuse to maintain Hb > 80.
- Perform bronchoscopy for suctioning of blood and secretions, assessment of bronchial anastomosis.
- Confirm position of DLETT and adequate cuff inflation.
- Confirm lung transplant ventilation settings (PCV 10, RR 10, PEEP 5 on room air (FiO2 21%)).
- Prepare Ambu bag for differential lung ventilation if first lung off ECMO.
- Draw up calcium, epinephrine, and fluid bolus in preparation for reperfusion.
- Slow, gradual recruitment under direct vision (max pressure 20–30 cmH_2O).
- Lung protective ventilation (+/− differential ventilation if first lung off ECMO).
- Gradual release of PA clamp over 10–15 min.
- Observe ECG for ST-segment changes, which may indicate coronary air emboli. The RCA is most affected due to its anterior position, resulting in inferior ST segment changes. This may require hemodynamic support to temporize.
- Repeat blood gas q10mins for first 30 min.

10.8 What Is the Management Algorithm for DLETT Obstruction and Poor Visibility during Bronchoscopy? (Table 10.4)

Table 10.4 Management of poor visibility on bronchoscopy

- Clean the end of the bronchoscope with sterile water, flush out the suction with sterile saline, and lubricate the scope as required.
- Confirm that the correct size of bronchoscope is being used. The adult bronchoscope will only fit through the lumen of a size 41 mm DLETT. All other DLETT sizes require the pediatric bronchoscope.
- The tube may have migrated during the surgery. Confirm tube depth and consider optimizing position under bronchoscopic guidance.
- Confirm with the surgeons that the bronchial anastomosis is free from external compression.
- Ensure that the bronchoscope light is turned on, white balance is calibrated, and suction is on and connected.
- It can be challenging to suction thick secretions through the small lumen of the pediatric bronchoscope. If you are having difficulty, consider using the endotracheal suction catheters that come with the DLETT. Be very careful when blindly passing a suction catheter to avoid disrupting the bronchial anastomosis.
- If thick or bloody secretions are not amenable to suction with the endotracheal suction catheter, consider performing pulmonary toilet with 5-10 cc of sterile saline to loosen the secretions.

10.9 How Is the Newly Implanted Lung Recruited and Ventilated?

Initial ventilation settings for all newly transplanted lungs should be: PCV 10, RR 10, PEEP 5–10, and FiO2 21%.

For transplants performed off ECMO, you may need to gradually increase vent settings to meet ventilatory requirements. Notify the surgeons if you need to increase vent settings over any of the following: PIP > 20 mmHg, RR > 20, PEEP >10, and FiO2 > 50%.

Recruitment may help optimize oxygenation but proceed cautiously. Use the "vital capacity" option on the ventilator to perform a controlled, sustained recruitment maneuver. Consider starting with a recruitment pressure of 20-30 mmHg for 15 s and increase only if significant portions of the lung remain collapsed.

10.10 What Is the Differential for Post-reperfusion Hypoxia? (Table 10.5)

Table 10.5 Causes of post-reperfusion hypoxia

Anesthesia and ventilation related
• Inadequate vent settings • Atelectasis/derecruitment • Tube obstruction or malposition • Volume overload and pulmonary edema • Transfusion associated lung injury (TRALI)
Surgical and transplant issues
• Primary graft dysfunction • Hyperacute rejection • Pulmonary venous anastomotic obstruction • Pulmonary arterial anastomotic obstruction • Bronchial anastomotic dehiscence • Torsion of allograft • Cardiac tamponade • Pulmonary embolism • Pre-existing PFO or other intracardiac shunt • Nitric oxide depletion after EVLP

10.11 How Is Post-reperfusion Hypoxia Managed? (Table 10.6)

Table 10.6 Management of post-reperfusion hypoxia

• Perform an ABG to assess severity of hypoxia, hypercarbia, acidosis. • Call for nitric oxide early, even as a precaution, as it can take up to 15 min for setup. • Inform surgeons of ventilatory issues and consider preparations for mechanical support if not already on ECMO. • Confirm vent settings. Increase PCV/PEEP/RR and perform recruitment as indicated. Temporize with 100% FiO2, especially if preparing to interrupt ventilation for bronchoscopy. • Bronchoscopy to suction secretions, identify obstructions, confirm tube placement, inspect anastomoses, and verify patency of bronchi. • Call for TEE if available to assess arterial and venous anastomoses, tamponade, embolism/thrombus, PFO. • Target MAP of 65-75 mmHg using vasopressors if required. • Trial of furosemide if patient hemodynamically stable and evident edema. • If no cause found, consider PGD as most likely cause. • Administer inhaled nitric oxide at 20-40 ppm. • Proceed with ECMO support (central VA if chest open, peripheral VV if chest closed)

10.12 What Are the Steps to Improve Poor Compliance in the Newly Transplanted Lung? (Table 10.7)

Table 10.7 Algorithm for poor lung compliance

- Confirm with the surgeons that the lungs and bronchi are free of external compression.
- Visual inspection by the surgeons can identify edema or areas of collapse.
- Ensure DLETT is not kinked. Bronchoscopy, suction, and toilet to remove any obstructive secretions or edema.
- If collapse is identified, perform a controlled recruitment maneuver (PIP 20–30 mmHg x 15 s) followed immediately by incremental increase in PEEP (2-3cmH20).
- Avoid increasing the I:E ratio which may temporarily improve oxygenation but at the expense of worsening lung injury [3].
- Consider diuresis if lungs are edematous or patient has significant positive fluid balance.
- Administer inhaled nitric oxide at 20-40 ppm.

10.13 What Is the Mechanism of Action of Inhaled Nitric Oxide (iNO)?

iNO is a selective pulmonary vasodilator with the potential to improve graft function through multiple mechanisms [4]:

1. Decreases pulmonary arterial resistance, reducing RV afterload.
2. Preferentially increases perfusion to well-ventilated lung regions, thereby enhancing oxygenation by optimizing the ventilation/perfusion ratio.
3. Prevents reperfusion edema by reducing pulmonary capillary pressure and inhibiting leukocyte-endothelial interactions.
4. Decreases ischemic-reperfusion injury and early inflammatory insults.
5. Prevents apoptosis.

10.14 What Is the Evidence Supporting the Use of iNO for Transplant?

Proof of concept was first demonstrated by Adatia et al. [5] in a small case series of 6 pediatric lung transplant recipients with either pulmonary hypertension or hypoxemia post-transplant [5]. They demonstrated that iNO lowered pulmonary artery pressure and pulmonary vascular resistance and improved intrapulmonary shunt fraction after lung transplantation.

In a landmark study by Date et al. [6] in 32 patients with immediate severe allograft dysfunction, iNO at a dose of 20-60 ppm was shown to improve oxygenation and decrease pulmonary artery pressure without systemic circulatory effects. They also reported shorter postoperative mechanical ventilation times, reduced airway complications, and mortality [6]. Similar favorable findings were reported by Ardehali et al. in a prospective trial involving 28 adult lung recipients [7].

Surprisingly, despite the major role of iNO in modern lung transplantation, there are relatively few human randomized controlled trials demonstrating improved oxygenation, reduced ischemia-reperfusion injury, or prevention of primary graft dysfunction [4, 8]. As a result, iNO has increasingly become scrutinized as an expensive but cosmetic treatment with only short-to-medium term benefits on gas exchange, without significant long-term benefit [9]. Notably, patients who are iNO-dependent at the end of surgery have poorer outcomes, with increased rates of grade 3 PGD and decreased 4-year survival [4]. These findings suggest that iNO dependency may be an early marker of poor prognosis.

10.15 Are There Any Negative Side Effects to iNO?

Inhaled NO has the potential to cause methemoglobinemia at higher doses (80-100 ppm) [10]. A recent study from Japan reported a NO-induced methemoglobinemia rate of 4%, even at doses of 10-20ppm [11]. Notably, iNO does not contribute to postoperative kidney injury in patients undergoing lung transplantation [12]. Although iNO has been shown to inhibit platelet aggregation and prolong bleeding time in neonates [13], there does not seem to be clinically significant effects on hemostasis in adults [14].

10.16 When Should iNO be Considered?

iNO should be considered post-reperfusion in the following scenarios:

- Persistent, refractory hypoxemia despite optimization of ventilatory parameters.
- Poor lung compliance, high airway pressures, or high FiO2 requirement which could compromise graft function.
- Persistent pulmonary hypertension.
- Moderate to severe RV dysfunction.

Be aware that in ARDS studies, approximately 1/3 of all ARDS patients are iNO non-responders who do not show significant improvements in oxygenation [15].

10.17 What Is the Effective iNO Dose in Lung Transplantation?

iNO dose-response studies have not been conducted for acute graft dysfunction post lung transplantation, as this remains an off-label indication. Dosing insights can be taken from the ARDS and cardiac surgery literature.

For patient with ARDS, a double-blind dose finding study and phase II trial randomized 177 patients to receive iNO at concentrations of 1.25, 5, 20, or 40ppm [16]. The percentage of patients in which iNO improved oxygenation (~60%) and the magnitude of change (increase in PaO2 > 20 mmHg) was not significantly different between groups. Furthermore, the effect of iNO on oxygenation disappeared in all groups by 48 h. The European expert recommendation on inhaled nitric oxide therapy in adults reflects these findings, suggesting that redistribution of blood flow and reduced intrapulmonary shunting can be achieved in most patients at concentrations ranging from 0.1—10 ppm, although they acknowledge that the optimum dose may vary over time and between different subjects [17]. This same guideline recommends an iNO dose of 10-20 ppm for acute vasodilator testing [17].

Conversely in patients with acute right heart failure secondary to respiratory disease, a dose-titration study from the cardiac surgery literature found that the mean concentration of iNO required to improve hemodynamics was 35 ppm. 85% of patients in their study who showed a substantial improvement in hemodynamics did so at a concentration less than 40ppm [18].

Incorporating these findings, it is reasonable to start iNO at 10-20 ppm for hypoxemia related to graft dysfunction, and 40 ppm for hemodynamic instability related to pulmonary hypertension or acute right heart failure.

10.18 Is There Utility in Prophylactic iNO?

Several studies have shown that iNO has limited benefit in patients without graft dysfunction [19, 20]. In a prospective, double-randomized trial of 33 recipients, the administration of 20 ppm prophylactic iNO at the time of reperfusion did not prevent pulmonary edema after lung transplantation, a sensitive marker for ischemia-reperfusion injury [20]. Anecdotally, iNO also significantly delays time to extubation, as abrupt weans can cause rebound pulmonary hypertension.

References

1. Moreno Garijo J, et al. The evolving role of extracorporeal membrane oxygenation in lung transplantation: implications for anesthetic management. J Cardiothorac Vasc Anesth. 2019;33(7):1995–2006.
2. Hoetzenecker K, et al. Intraoperative extracorporeal membrane oxygenation and the possibility of postoperative prolongation improve survival in bilateral lung transplantation. J Thorac Cardiovasc Surg. 2018;155(5):2193–206 e3.
3. Muller-Redetzky HC, et al. Increasing the inspiratory time and I:E ratio during mechanical ventilation aggravates ventilator-induced lung injury in mice. Crit Care. 2015;19:23.
4. Fessler J, et al. Inhaled nitric oxide dependency at the end of double-lung transplantation: a boosted propensity score cohort analysis. Transpl Int. 2019;32(3):244–56.
5. Adatia I, et al. Inhaled nitric oxide in the treatment of postoperative graft dysfunction after lung transplantation. Ann Thorac Surg. 1994;57(5):1311–8.
6. Date H, et al. Inhaled nitric oxide reduces human lung allograft dysfunction. J Thorac Cardiovasc Surg. 1996;111(5):913–9.
7. Ardehali A, et al. A prospective trial of inhaled nitric oxide in clinical lung transplantation. Transplantation. 2001;72(1):112–5.
8. Ramadan ME, et al. Con: inhaled pulmonary vasodilators are not indicated in patients undergoing lung transplantation. J Cardiothorac Vasc Anesth. 2017;31(3):1127–31.

9. Glanville AR. Inhaled nitric oxide after lung transplantation: no more cosmesis? Am J Respir Crit Care Med. 2003;167(11):1463–4.
10. Weinberger B, et al. The toxicology of inhaled nitric oxide. Toxicol Sci. 2001;59(1):5–16.
11. Yoshiyasu N, et al. Current status of inhaled nitric oxide therapy for lung transplantation in Japan: a nationwide survey. Gen Thorac Cardiovasc Surg. 2021;69(10):1421–31.
12. Ri HS, et al. Inhaled nitric oxide therapy was not associated with postoperative acute kidney injury in patients undergoing lung transplantation: a retrospective pilot study. Medicine (Baltimore). 2018;97(22):e10915.
13. George TN, et al. The effect of inhaled nitric oxide therapy on bleeding time and platelet aggregation in neonates. J Pediatr. 1998;132(4):731–4.
14. Goldstein B, et al. Effects of inhaled nitric oxide on hemostasis in healthy adults treated with heparin: a randomized, controlled, blinded crossover study. Thromb J. 2012;10:1.
15. Rossaint R, et al. Our paper 20 years later: Inhaled nitric oxide for the acute respiratory distress syndrome--discovery, current understanding, and focussed targets of future applications. Intensive Care Med 2014;40(11):1649–1658.
16. Dellinger RP, et al. Effects of inhaled nitric oxide in patients with acute respiratory distress syndrome: results of a randomized phase II trial. Inhaled nitric oxide in ARDS study group. Crit Care Med. 1998;26(1):15–23.
17. Germann P, et al. Inhaled nitric oxide therapy in adults: European expert recommendations. Intensive Care Med. 2005;31(8):1029–41.
18. Bhorade S, et al. Response to inhaled nitric oxide in patients with acute right heart syndrome. Am J Respir Crit Care Med. 1999;159(2):571–9.
19. Tavare AN, Tsakok T. Does prophylactic inhaled nitric oxide reduce morbidity and mortality after lung transplantation? Interact Cardiovasc Thorac Surg. 2011;13(5):516–20.
20. Perrin G, et al. Inhaled nitric oxide does not prevent pulmonary edema after lung transplantation measured by lung water content: a randomized clinical study. Chest. 2006;129(4):1024–30.

Hemodynamics and ECMO

11

Abstract

Hemodynamic management in lung transplantation centers on maintaining adequate perfusion while avoiding elevations in pulmonary artery pressure. Specific hemodynamic goals depend on whether the surgery is performed on or off ECMO. Off ECMO aims for MAP 60–70 mmHg, raising to 80–90 mmHg in patients with vascular disease. Pulmonary artery (PA) test clamping identifies high PA pressures and RV vulnerability. Intraoperative ECMO is preemptively initiated in cases of severe PH, RV dysfunction, or challenging anatomy. On ECMO targets MAP 60–80 mmHg, with flows adjusted for oxygenation, perfusion, and CO2 clearance. Loss of PA pulsatility suggests hypovolemia or excessive ECMO flow.

11.1 What Are the Hemodynamic Targets off ECMO?

Ideally, maintain blood pressures within 20% of the patient's baseline. A more practical target would be to maintain a MAP of 60-70 mmHg given the changes that can occur with mediastinal retraction. In the patient with comorbid cardiovascular disease (coronary artery disease, stroke), consider raising the MAP target to maintain perfusion through disease vessels (MAP 80–90 mmHg). As mediastinal retraction increases with each subsequent anastomosis, target a higher MAP as transient hypertension is less harmful than recurrent or prolonged hypotension. Targeting higher pressures also protects against the hypotension that can accompany reperfusion.

M. Salvatori, A. Huang (eds.), *Anesthesia for Lung Transplant*,
https://doi.org/10.1007/978-3-032-06112-6_11

11.2 What Is the Purpose of the PA Test Clamp?

Following dissection of the native lung from the mediastinum and liberation of the main PA, pulmonary veins, and bronchus, a test clamp is applied on the PA. The purpose of the test clamp is to assess for hemodynamic instability or high PA pressures prior to committing to a pneumonectomy [2]. Significant increase in PA pressure should prompt discussion about mitigating strategies including iNO, or more likely, the use of ECMO. This is especially important during the test clamp of the second lung, as high PA pressures can result in lung injury and allograft dysfunction [3, 4].

11.3 What Are the Signs of Impending RV Failure?

Patients with pulmonary hypertension and/or baseline cardiac dysfunction may decompensate intraoperatively due to the stresses of surgery and anesthesia. Factors that may precipitate acute RV failure include induction of anesthesia, hypoxia/hypercarbia, positive pressure ventilation, sustained hypotension, large or rapid fluid boluses, acidosis, surgical stimulation and/or manipulation, light anesthesia, and PA test clamping.

Hemodynamic signs of worsening RV dysfunction include severe and refractory hypotension and a rising CVP. If the PA catheter has been withdrawn to the RV, the trace may show the "square root sign" indicative of impaired RV relaxation and an elevated RV end diastolic pressure (Table 11.1). A late but ominous sign is a rapid drop in the PA pressures, which suggests that the RV can no longer generate the force required.

These hemodynamic signs are subtle and can be easily missed. We therefore recommend the use of TEE for all patients with near-systemic pulmonary pressures in order to continuously evaluate RV function throughout surgery. TEE signs of RV dysfunction include acute RV dilatation, leftward bowing of the interatrial septum, acute worsening of tricuspid regurgitation, a fractional area change <35% and a TAPSE <1.7 cm. Seeing any of these signs should prompt a discussion about completing the transplant on ECMO.

11.4 Why Does Clamping the Atrium Affect Hemodynamics?

The order of implantation is main bronchus, PA, and then left atrial cuff anastomosis. During this final stage, partial clamping of the atrium is required to size the atrial cuff and complete the anastomosis. Clamping may result in dysrhythmias from manipulation, or hypotension and increased CVP due to decreased venous return from retraction of the heart. Clamping may also inadvertently occlude the circumflex artery, resulting in myocardial ischemia.

Table 11.1 Differential and PAC changes for high PA pressures

Elevated RV Systolic Pressure Pulmonary hypertension Pulmonic valve stenosis Significant ASD or VSD Pulmonary embolism Elevated RV End Diastolic Pressure	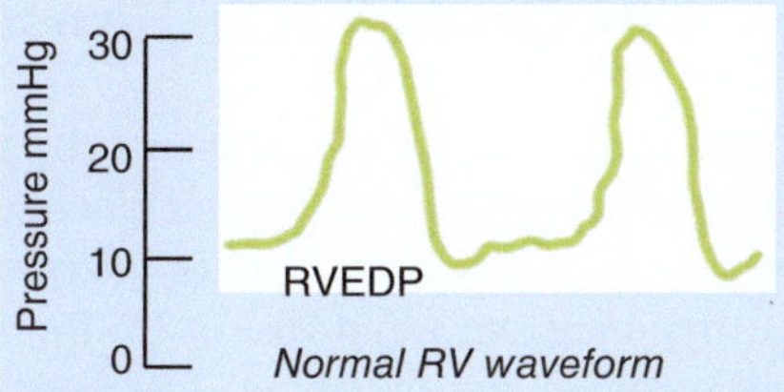
Cardiomyopathy RV ischemia Cardiac constriction Cardiac tamponade Square root sign Restrictive cardiomyopathy Cardiac tamponade	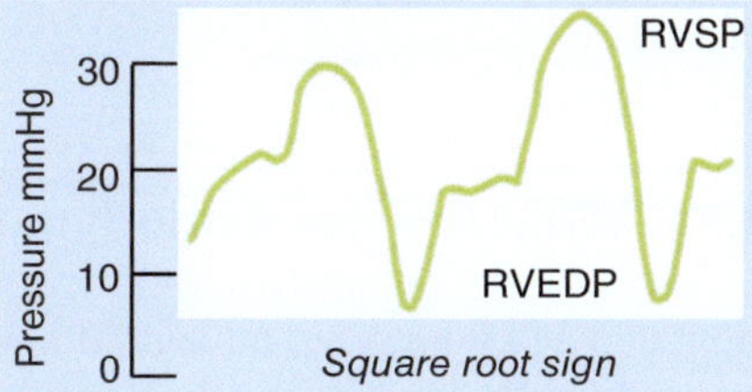

A pathological RV waveform is the "square root sign". Early diastolic ventricular filling is abnormally rapid, but late diastolic filling halts abruptly when total cardiac volume reaches the limit set by the stiff ventricle or pericardium. ASD, atrial septal defect; RV, right ventricle; RVEDP, right ventricular end diastolic pressure; RVSP, right ventricular systolic pressure; VSD, ventricular septal defect.

11.5 When Do the Surgeons Pre-emptively Decide to Perform the Lung Transplant on Elective ECMO?

There are several scenarios in which the surgeon may decide to electively perform the transplant on ECMO.

- Patients bridged to transplant with extracorporeal support are circuit-dependent and will always require ECMO support for transplantation. Although transplants can be performed on V-V ECMO (**See Question 17.3**), the surgeons may convert to a central V-A ECMO configuration to better support hemodynamics.
- ECMO is routinely used for patients with severe pulmonary hypertension and/or right ventricular dysfunction. This can be instituted pre-induction (peripherally) or early after clamshell depending on clinical status.
- If there are any concerns regarding graft quality or ischemic time, ECMO can be used to transplant both lungs before relying on for either for oxygenation. ECMO also allows for more controlled reperfusion, minimizing risk of reperfusion injury.
- ECMO is often used for patients with known anatomical abnormalities, dissection challenges, or small hemithoraces, as adequate exposure may necessitate aggressive retraction causing hemodynamic instability.
- Lobar transplants are a common indication for early/planned ECMO, due to the small hemithoraces and need to avoid hyperperfusion of the transplanted lobe.

11.6 When Should Emergency Conversion to ECMO Be Considered for Hemodynamic Reasons?

ECMO is indicated if PA pressures rise to near systemic pressures with test clamp, causing right ventricular dysfunction and/or hemodynamic instability. RV dysfunction may be detected on TEE, or interpreted from changing hemodynamic parameters (hypotension, rapid rise in CVP). The need for ECMO may arise only after reperfusion of the first lung if systolic PA pressure rises to >50 mmHg on test clamping of the second lung.

11.7 What Are the Hemodynamic Targets on ECMO?

Maintain a MAP between 60 and 80 mmHg on V-A ECMO [1]. An elevated MAP (i.e., > 90 mmHg) may impede ECMO flows (high SVR) and increase the risk of aortic dissection or stroke. Low MAP may compromise end-organ perfusion. Persistent MAP above 90 mmHg or below 50 mmHg should prompt discussion with the surgical team.

11.8 How Is Hypertension at the Time of Cannulation Managed?

The target blood pressure for aortic cannulation is a systolic BP between 80 and 100 mmHg. SBP >100 mmHg increases risk of dissection, whereas SBP < 80 mmHg increases risk of injury to the posterior aortic wall. Common pharmacologic options include nitroglycerin (40 μg boluses), magnesium sulfate (1g boluses), propofol, or IV opioids. Reverse Trendelenburg positioning is effective and reversible but may disrupt the surgeons at a critical part of the procedure.

11.9 What Is the Significance of the ECMO Flow Rate?

On central V-A ECMO, venous blood is drawn from the right atrium and returned to the ascending aorta. The heart and lungs are bypassed by the ECMO circuit, which provides oxygenation, gas exchange, and forward flow. However, V-A ECMO is not an "on or off" system. The ECMO circuit and native cardiopulmonary circulation exist in parallel. The "full flow" rate is relative, based on a formula that calculates cardiac index using body surface area (BSA). The more blood that flows through the ECMO circuit, the less blood traverses the heart and lungs.

There are 3 principal benefits to V-A ECMO in lungs transplant:

1. Oxygenation is maintained independent of lung function
2. Blood flow and pressure through the native cardio-pulmonary circulation can be finely controlled
3. Assisted cardiac output

ECMO flow rate may therefore require adjustment for ventilatory, hemodynamic, and metabolic reasons:

Ventilation ECMO flows need to be increased during periods of apnea of impaired native lung function. Remember to increase ECMO to full flow to compensate for prolonged apnea during tube exchange.

Hemodynamics Increases in ECMO flows may be required to support blood pressure or temporize instability from surgical retraction. ECMO flow may need to be reduced to temporize hypertensive spikes, or to allow partial blood flow through the pulmonary circulation to the new graft **(See Question 11.13)**.

Metabolics High lactate and low SvO_2 indicate inadequate oxygen delivery (oxygen delivery (DO_2) = cardiac output (CO) × arterial oxygen content (CaO_2)). Strategies such as correcting anemia with transfusion (to improve oxygen carrying capacity), increasing ECMO flows (to optimize MAP/perfusion pressure), and adjustment of the circuit FiO_2 (increase oxygen content) should be considered.

11.10 How Is Blood Pressure Managed on ECMO?

The most common causes of hypotension on ECMO are hypovolemia, vasoplegia, mechanical compression, and inadequate ECMO flows. All these issues typically respond to volume, so administer 250-500 cc boluses, repeating as indicated while assessing fluid responsiveness. After 2-3 L of crystalloid, consider introducing 5% albumin, which may improve microcirculation, oxygen delivery, and spare larger volumes of crystalloid [1] (**See Question 14.3**). If the patient is not fluid responsive, use pressors (norepinephrine, vasopressin, phenylephrine) to address vasoplegia. The use of pressors to increase SVR must be weighed against the risks of increased afterload and LV stroke work.

MAP is a function of both cardiac output and systemic vascular resistance (MAP = CO x systemic vascular resistance (SVR)). On ECMO, both native cardiac ejection and ECMO flows contribute to the overall “cardiac output”. Inotropy and augmentation of cardiac contractility are not commonly required during lung transplantation. However, the perfusionist may increase ECMO flows, thereby increasing the ECMO contribution to cardiac output (**See Question 11.9**). If the patient is volume replete but full ECMO flows cannot be achieved, the surgeons may need to adjust cannula position.

Hypertension typically responds to anesthesia and analgesia. If the hypertension is extreme (MAP >100 mm Hg for >5 minutes) and refractory, the perfusionist can temporize by reducing ECMO flows. Very rarely, vasodilators such as nitroglycerin or nitroprusside are required. Recognize that high doses of vasodilators may impair hypoxic pulmonary vasoconstriction (HPV), leading to shunt and worsening oxygenation.

11.11 What Is Sweep and When Should It Be Adjusted?

Sweep refers to gas flow through the oxygenator that removes CO_2 from blood. On V-A ECMO, the sweep gas is 50–100% oxygen, titrated to flow rate and PaO_2. The amount of CO_2 removed depends on sweep gas flow rate, independent of blood flow. Sweep gas flow rates are typically between 0-5 L/min and are adjusted in 0.5 L/min increments to target a $PaCO_2$ of 40-45 mmHg.

In COPD patients with chronic CO_2 retention, aggressive normalization of $PaCO_2$ should be avoided to allow time for metabolic adaptation. In these cases, the perfusionist will often titrate sweep to normalize pH.

Sweep rates should remain stable or lessen over the course of the transplant. Once the first lung is reperfused, ECMO flows are reduced to allow blood flow through the PA. The fraction of blood allowed to pass through the native cardiopulmonary circulation will not have CO_2 clearance via ECMO, and ventilator settings need to be adjusted accordingly. $EtCO_2$ tracings may be helpful in adjusting ventilatory parameters, however dead space can be variable. Although sweep rates can be adjusted to eliminate this residual CO2, attempt to manage by gradually increasing ventilation, as sweep will not be available once ECMO is discontinued.

11.12 When Should the Surgeons Be Notified of a High CVP on ECMO?

Notify the surgeons if the CVP climbs to ~20 mmHg or greater for over 5 min. An abrupt and/or persistent elevation in CVP is concerning for excessive cardiac retraction, or development/worsening of pulmonary hypertension and/or RV failure.

11.13 What Is the Significance of PA Pulsatility?

With reperfusion of the first lung, ECMO flows are reduced to 50–60% to allow blood to flow through the pulmonary artery, evidenced by PA pulsatility. The bronchial circulation is not commonly restored during transplantation, and blood supply to the new lungs is derived solely from the PA blood flow. PA blood flow and pulsatility is essential to graft perfusion and may reduce the risk of PGD.

Loss of PA pulsatility likely reflects hypovolemia, mechanical compression, or high ECMO flows. Administer 250-500 cc bolus and reduce ECMO flows as tolerated, allowing more blood to pass through the native cardiopulmonary circulation. If the PA waveform remains flat despite these interventions, discuss temporary relief of mechanical retraction with the surgical team.

Rarely, loss of PA waveform may result from iatrogenic damage to the PA catheter during pneumonectomy, marked by the appearance of blood in the balloon syringe or thermistor. If damage is suspected, the PA catheter must be removed or replaced at the end of the case.

Table 11.2 Comparison of ECMO types

	V-V ECMO	V-A ECMO
Indication	Respiratory failure	Cardiogenic shock
Drainage cannula	IVC, SVC	IVC, RA
Return cannula	RA	Aorta
Hemodynamic support	NO	YES
Bypasses heart + lungs	NO	YES
Pulsatile flow	Preserved	Reduced
Cannulation	Peripheral	Peripheral or central

11.14 When Should ECMO Be Continued Post-transplant?

ECMO should be continued after transplantation to bridge patients with early graft dysfunction, hemodynamic, instability, or pulmonary hypertension. If the transplant is performed on central V-A ECMO, and decompensation occurs prior to decannulation or chest closure, the surgeons may choose to transfer the patient to the ICU centrally cannulated. This is a rare exception, as it warrants a return to the OR for decannulation once conditions improve. Maintaining central V-A ECMO is more likely when the indication is hemodynamic compromise, pulmonary hypertension, or ventricular dysfunction.

More commonly, post-transplant ECMO is required for early primary graft dysfunction resulting in inadequate oxygenation or ventilation. In these cases, the surgeons will convert the patients from central V-A ECMO to peripheral V-V ECMO using the femoral or internal jugular veins. This allows ventilatory support with minimal impact on the cardiovascular system and allows for easy bedside removal in ICU without a return to the OR.

Conversion from central to peripheral V-A ECMO is rare as it puts undue strain on the heart, as the LV must pump against the high afterload generated by the arterial return cannula. If peripheral V-A ECMO is performed as rescue in the ICU, a venting strategy such an atrial septostomy or percutaneous Impella is required to decompress the LV. If the LV is not adequately decompressed, high LVEDP and LA pressures may result in refractory pulmonary edema of the newly transplanted lungs. See Table 11.2 for comparison of ECMO types.

11.15 What Causes ECMO Chatter?

Drainage insufficiency occurs when there is decreased ECMO flow due to inadequate pump preload. Intermittent negative pressure in the circuit causes rhythmic swinging of the drainage cannula, a phenomenon termed “chatter” or “chugging” [5]. The perfusionist alerts the team when chatter is persistent and significantly impairs ECMO flows. Table 11.3 lists possible contributors.

Table 11.3 Differential for ECMO chatter

• Hypovolemia (aggressive pre-op diuresis, insensible losses, occult bleeding)
• Cannula malposition or obstruction
• Spontaneous ventilation attempts or coughing
• Vasodilation
• Excessive pump speeds
• Abdominal compartment syndrome

References

1. Moreno Garijo J, et al. The evolving role of extracorporeal membrane oxygenation in lung transplantation: implications for anesthetic management. J Cardiothorac Vasc Anesth. 2019;33(7):1995–2006.
2. Gust L, et al. Single-lung and double-lung transplantation: technique and tips. J Thorac Dis. 2018;10(4):2508–18.
3. Fullerton DA, et al. Pulmonary vasomotor dysfunction is produced with chronically high pulmonary blood flow. J Thorac Cardiovasc Surg. 1996;111(1):190–7.
4. Lick SD, et al. Technique of controlled reperfusion of the transplanted lung in humans. Ann Thorac Surg. 2000;69(3):910–2.
5. Zakhary B, et al. How I manage drainage insufficiency on extracorporeal membrane oxygenation. Crit Care. 2020;24(1):151.

12 Reperfusion

Abstract

Reperfusion is often the most challenging aspect of solid organ transplant, as the anesthesiologist must simultaneously manage the accompanying hemodynamic, ventilatory, and metabolic derangements. Whether on or off ECMO, reperfusion requires precise coordination between surgical and anesthetic teams. Reperfusion begins with controlled pulmonary artery unclamping to minimize the risk of reperfusion injury. Recruitment maneuvers and lung-protective ventilation at low FiO_2 optimize graft compliance and minimize injury. Ventilation strategies are tailored per graft performance, with iNO used as needed. On ECMO, gas exchange is supported, allowing gentler reperfusion and gradual weaning. Associated risks include hypotension, hypoxia, coronary air embolism, and early graft dysfunction. Close monitoring, frequent ABGs, and readiness to escalate to ECMO are essential for safe reperfusion and graft function.

12.1 How Is Lung Reperfusion Managed off ECMO?

When all anastomoses are complete and hemostasis is achieved, the surgeons will gradually reperfuse the lung and request ventilation (See Table 10.3 for pre-reperfusion checklist). Having two anesthesia providers is beneficial during reperfusion "off ECMO" as this is a dynamic and complex process requiring expedient and coordinated management. Set the FiO2 to room air, increasing the flow rate to ensure the target FiO_2 is reached quickly. Begin initially with a recruitment maneuver of 20cmH_2O for 5–10 s and repeat as indicated. Observe the lung to ensure complete inflation while avoiding overdistention. Recruitment is essential to optimizing compliance of the newly implanted graft. After complete recruitment, ventilate with pressure control settings of 10, rate of 10, and PEEP of 5–10. A tidal

M. Salvatori, A. Huang (eds.), *Anesthesia for Lung Transplant*,
https://doi.org/10.1007/978-3-032-06112-6_12

volume of about 200–250 cc should be targeted, and overinflation avoided. Communicate the ventilator settings, compliance, and tidal volumes to the surgeons, especially if changes are required. The surgeon may request intermittent deflation to assess or correct any anastomosis or bleeding.

At the same time, manually ventilate the remaining native lung with an Ambu Bag connected to the auxiliary O2 flowmeter. Tidal volume and rate should be high to compensate for the newly perfused lung. Tidal volume and rate are often inconsistent with manual ventilation, so repeat ABGs q10 mins to ensure adequate oxygenation and gas exchange.

Restoration of blood flow to the newly implanted lung can cause transient hypotension due to relative hypovolemia. Pre-emptive administration of 300–500 cc of IV fluid approximately 5 min before reperfusion can lessen hemodynamic fluctuations. The PA clamp is opened slowly over 5–10 min which limits the rate of reperfusion and allows for rapid reclamping should bleeding occur [1]. Animal studies demonstrate that abrupt clamp removal with full instantaneous reperfusion increases the risk of reperfusion injury [2–4].

With the newly implanted lung completely reperfused and ventilatory parameters stable, proceed with a trial of OLV using the newly implanted graft in close communication with the surgical team. Increase vent settings and FiO2 to maintain adequate oxygenation and ventilation. Notify the surgeons if you need to increase vent settings over any of the following: PIP > 20 mmHg, RR > 20, PEEP >10, FiO2 > 50%.

As with reperfusion of the first lung, the second newly implanted lung should be recruited and ventilated with lung protective settings and the lowest FiO_2 possible. Ventilation of the first graft during reperfusion of the second can be challenging, and largely depends on graft performance. If adequate one lung oxygenation and ventilation were achievable with low FiO_2 (30–40%) during transplantation of the second lung, then consider returning to two-lung ventilation at a relatively low FiO_2 (30–40%) and lung protective vent settings. Target a Vt of 4-6 cc/kg, avoiding peak airway pressures >25cmH20. Adjust respiratory rate to target a PaCO2 of 40–45 mmHg and optimize compliance with PEEP of 5-10cmH20.

If performance of the first graft is marginal and oxygen requirements remain high (i.e., FiO2 > 50%), discuss optimization strategies with the surgical team prior to reperfusion. Management may include recruitment, bronchoscopy, or iNO. Lack of improvement necessitates manual ventilation of the first lung with an Ambu-Bag and 100% FiO_2 during the 5 to 10-min period of reperfusion. The first graft will likely tolerate this short period of hyperoxygenation without long-term effects. Return to two lung ventilation with intermediate FiO2 as soon as tolerated. If the quality of the second graft is in question, conversion to ECMO may be required.

12.2 How Is Lung Reperfusion Managed on ECMO?

On ECMO, reperfusion of the first lung follows the same sequence described off ECMO, including an initial recruitment maneuver. As the ECMO circuit supports oxygenation, both lungs remain on the ventilator with lung protective settings and low FiO_2. On V-A ECMO, reperfusion can be controlled via the ECMO flow rate, and the PA clamp is opened more quickly. On V-V ECMO, the PA clamp requires gradual opening as per off ECMO transplants. The same sequence is repeated for the second lung, including recruitment maneuver, low FiO_2, and lung protective settings. The ECMO circuit supports oxygenation during this period and is gradually weaned as tolerated.

12.3 What Are the Major Risks Associated with Reperfusion?

Reperfusion is a dynamic process that involves coordination between the surgical and anesthetic teams, as a series of both hemodynamic and ventilatory changes are occurring. During reperfusion, hypotension may result from:

1. Blood loss at the PA or PV anastomoses
2. Systemic circulation of ischemic metabolites
3. Myocardial ischemia from air emboli to the RCA
4. Expansion of the patient's circulation volume
5. Ongoing mediastinal/surgical retraction for exposure

In anticipation of these changes, administer 300–500 cc of IV fluid 5 min prior to reperfusion. At the time of reperfusion, bolus vasopressors and/or increase infusions as required. If coronary air embolism is suspected from inferior ST changes, RV dysfunction, hypotension, or acute rise in CVP, epinephrine boluses (4–8ug) can help temporize the transient instability.

Hypoxia and increasing vent requirements are common post-reperfusion, or after liberation from ECMO. Intermittent OLV is often employed to address anastomotic concerns, resulting in shunting with inadequate time for compensatory HPV. Even when ventilating both lungs, early graft performance may be poor, as gas exchange typically improves over hours to days. Rarely, early reperfusion injury, PGD, and pulmonary edema can occur, further compromising gas exchange. Relative to other solid organ transplants, metabolic derangements during reperfusion are mild. It is uncommon to treat metabolic acidosis or hyperkalemia during reperfusion.

References

1. Nguyen DC, et al. Review of donor and recipient surgical procedures in lung transplantation. J Thorac Dis. 2019;11(Suppl 14):S1810–S16.
2. Pierre AF, et al. Rapid reperfusion causes stress failure in ischemic rat lungs. J Thorac Cardiovasc Surg. 1998;116(6):932–42.

3. Bhabra MS, et al. Controlled reperfusion protects lung grafts during a transient early increase in permeability. Ann Thorac Surg. 1998;65(1):187–92.
4. Halldorsson A, et al. Controlled reperfusion after lung ischemia: implications for improved function after lung transplantation. J Thorac Cardiovasc Surg. 1998;115(2):415–24; discussion 24–5.

Primary Graft Dysfunction 13

Abstract

Primary graft dysfunction is a syndrome of acute lung injury following lung transplantation, defined by hypoxemia and pulmonary edema on imaging. The ISHLT classifies PGD by PaO_2/FiO_2 ratio and chest X-ray findings. PGD is largely driven by ischemia-reperfusion injury and inflammatory processes. Risk factors span donor, recipient, and intraoperative variables, including single-lung transplant, high reperfusion FiO_2, and intraoperative transfusion. Management parallels ARDS strategies: lung-protective ventilation, fluid restriction, and in severe cases, ECMO support. Prevention focuses on optimal donor selection, improved preservation techniques, and meticulous intraoperative management including controlled reperfusion and early escalation to extracorporeal support.

13.1 What Is Primary Graft Dysfunction?

Primary graft dysfunction (PGD) is a syndrome of acute lung injury occurring early after lung transplantation [3]. The consensus-based definition from the International Society for Heart and Lung Transplantation (ISHLT) is based on diffuse alveolar infiltrates on chest X-ray and hypoxemia as evidenced by the PaO2/FiO2 (PF) ratio (Table 13.1). In the operating room, lungs with PGD are characteristically edematous with poor compliance and gas exchange [4]. PGD Grade 0 denotes the absence of acute lung injury. Note that the presence of edema on chest X-ray is both necessary and sufficient. Patients without alveolar infiltrates on chest imaging are graded as PGD 0 regardless of their P/F ratio [5]. PaO2/FIO2 should ideally be measured on PEEP of 5cmH2O at FIO2 of 1.0 while patients are on mechanical ventilation; however, grading should not be altered for other settings. Notably, the use of nitric oxide or other pharmacologic agents to improve oxygenation does not affect

M. Salvatori, A. Huang (eds.), *Anesthesia for Lung Transplant*,
https://doi.org/10.1007/978-3-032-06112-6_13

Table 13.1 2016 ISHLT PGD definition [3]

Grade	Pulmonary edema on CXR	PaO2/FiO2 ratio
PGD Grade 0	No	Any
PGD Grade 1	Yes	>300
PGD Grade 2	Yes	200–300
PGD Grade 3	Yes	<200

grading. If the patient is on extracorporeal lung support (ECLS) and has bilateral pulmonary edema on chest X-ray, PGD is defined as grade 3 and ECLS use should be recorded [3].

At TGH, our PGD3 rates are 26% of patients at 0 h, and 19.2% of patients at 72 hours, which is comparable with the rates published in literature [6]. Single lung transplants have a higher incidence of PGD3 at all time points.

Several epidemiological studies have shown increased 30-day and 1-year mortality in patients with PGD, with the presence of PGD3 being the best predictor of long-term mortality. Prekker et al. demonstrated that T0 PGD data were the most complete, whereas arterial blood gases were not measured at T72 in most of their patients. However, the authors found that long-term mortality was higher based on the T24 or T48 time points and the presence of PGD grade 3, rather than T0 or the PGD grade 1. Subsequent studies likewise have suggested that T48 and T72 best capture the most clinically significant forms of PGD. 2,24 Furthermore, using latent class analyses, Shah et al. described that patients with T0 PGD3 who continued to have PGD3 at T48 and T72 had the greatest risk of mortality compared with those patients with PGD3 only at earlier time points.

Because of high data completeness, T24 may have robustness and allow assessment of the effect of perioperative PGD management strategies; however, PGD3 present at later times (T48 and T72) appears to have the greatest effect on long-term outcomes, including BOS and mortality. Thus, PGD3 present at T48 and/or T72 seems to be most suited for use as an outcome in clinical trials because of its clinical impact [1, 7–12]. A retrospective cohort study showed 30-day mortality rates of 3.5% for PGD grade 1, 6.2% for grade 2, and 24.5% for PGD grade 3 at 24 hours [13]. Of note, single lung transplants show less robust correlation between PGD 3 and outcomes.

13.2 What Is the Pathophysiology of PGD?

PGD is a complex and multifactorial syndrome without a singular, proven etiology. PGD likely reflects the culmination of several insults inherent to lung transplantation, including mechanical, immune, inflammatory, and microbial factors [3]. Although preservation techniques have evolved to minimize the risk, ischemic-reperfusion injury (IRI) remains a leading cause of primary graft dysfunction in lung transplantation. IRI occurs after prolonged interruption in organ perfusion, followed by restoration of blood flow. IRI affects all solid organ transplantation and is a major source of morbidity and mortality, particularly in lung transplants. IRI is characterized by sterile inflammation, microvascular permeability, endothelial cell dysfunction, and pulmonary edema with increased pulmonary vascular resistance and impaired oxygen exchange [14]. Aggravating factors include cold ischemic storage, mechanical ventilation, pneumonia, aspiration, brain death, hypotension, and trauma, which serve to amplify the activation of inflammatory parameters [15]. Although a full discussion of inflammatory parameters is beyond the scope of this text, a concise review has been published by the ISHLT [4].

13.3 What Are the Risk Factors for PGD? (Table 13.2)

Table 13.2 PGD risk factors [6]

Recipient	Donor	Operative
• Etiology of lung disease: COPD 11.8% CF 12.4% IPF 18% IPAH 30% Sarcoidosis 50% • Obesity (2x) • Baseline elevation in inflammatory markers • Markers of airway epithelial injury (CC16) • Innate immunity activation • Genetic prostaglandin polymorphisms	Definite: • Cigarette smoking	Definite: None
	Probable: • Aspiration • Chest trauma/pulmonary contusion • Undersized donor relative to recipient • Heavy alcohol use • Extended criteria donors	Probable: • Previous pleurodesis • Single lung transplant • Large volume intraoperative blood product transfusion • Reperfusion FiO2 > 40% • Undersized lungs
	Possible: • Age • Oxygenation • Chest X-ray abnormalities • Purulent secretions • Thromboembolism and fat embolism • Brain death • Pre-existing inflammatory status • Traumatic brain injury • Time from brain death to cold preservation • Prolonged mechanical ventilation	Possible: • CPB vs. ECMO • Ischemic time • Delayed chest closure
Notable non-contributors		
• Age • Sex	• Gender • Donor-recipient gender mismatch • Race • Positive sputum gram stain • Donor P/F ratio	• Prior cardiothoracic surgery

13.4 How Is PGD Managed?

Although there is no clear consensus on management of PGD, the core principles of acute respiratory distress syndrome (ARDS) are broadly applicable. The primary goal of therapy is to support ventilation during lung recovery while minimizing oxygen toxicity, volutrauma, and barotrauma [2]. Fluid restriction and low tidal volume ventilation with PEEP are the mainstays of treatment. Although iNO and inhaled prostaglandins have been shown to reduce pulmonary vascular resistance and VQ mismatch, long-term RCT outcome data regarding the use of pulmonary vasodilators in lung transplantation is lacking. Novel therapies for PGD include surfactant, complement inhibition, platelet-activating factor antagonists, platelet and neutrophil traps, stem cells, and plasmapheresis [16]. In severe or refractory cases, continued post-transplant ECMO support may be required to support oxygenation and ventilation until the lungs recover. If all these management strategies fail, retransplantation remains the only viable option, with a TLTP rate of approximately 5%.

13.5 How Is the Risk of PGD Minimized?

The ISHLT highlights the 3 strategies for mitigating the risk of PGD [16]:

1. Optimizing donor and recipient selection
    - Thorough evaluation encompassing gas exchange, chest radiography, bronchoscopy, and potentially biomarkers
    - Declining lungs at high risk of acute lung injury (ALI) before or after brain death
    - Donor management including hemodynamics, hormonal resuscitation, electrolytes and fluid control and body temperature maintenance
    - Respiratory optimization pre-procurement including high levels of PEEP (15 cmH2O); inspiratory pressures of 25 cmH2O; bronchoscopy to assess and minimize respiratory secretions; 30° head elevation; and endotracheal cuff pressures of 25 cmH2O to limit aspiration
    - Reconditioning with EVLP
2. Improving lung preservation storage techniques (ISHLT protocol [16])
3. Improving lung implantation and reperfusion techniques
    - Cooling during implantation
    - Intraoperative extracorporeal support
    - Avoidance of blood transfusion wherever possible
    - Avoid use of cardiotomy sucker in full CPB, prioritize cell-saver use
    - Controlled reperfusion via slow PA clamp release
    - If performed on ECMO, maintain perfusion, PA pulsatility, and ventilation to the newly implanted lung while the second lung is being implanted

- Lung protective ventilation of the newly implanted lungs
- Prophylactic post-operative ECLS

References

1. Diamond JM, et al. Clinical risk factors for primary graft dysfunction after lung transplantation. Am J Respir Crit Care Med. 2013;187(5):527–34.
2. Di Nardo M, et al. Postoperative management of lung transplant recipients in the intensive care unit. Anesthesiology. 2022;136(3):482–99.
3. Snell GI, et al. Report of the ISHLT working group on primary lung graft dysfunction, part I: definition and grading-a 2016 consensus group statement of the international society for heart and lung transplantation. J Heart Lung Transplant. 2017;36(10):1097–103.
4. Gelman AE, et al. Report of the ISHLT working group on primary lung graft dysfunction part III: mechanisms: a 2016 consensus group statement of the international society for heart and lung transplantation. J Heart Lung Transplant. 2017;36(10):1114–20.
5. Christie J, et al. Potential refinements of the international society for heart and lung transplantation primary graft dysfunction grading system. J Heart Lung Transplant. 2008;27(1):138.
6. Diamond JM, et al. Report of the international society for heart and lung transplantation working group on primary lung graft dysfunction, part II: epidemiology, risk factors, and outcomes-a 2016 consensus group statement of the international society for heart and lung transplantation. J Heart Lung Transplant. 2017;36(10):1104–13.
7. Christie JD, et al. The effect of primary graft dysfunction on survival after lung transplantation. Am J Respir Crit Care Med. 2005;171(11):1312–6.
8. Daud SA, et al. Impact of immediate primary lung allograft dysfunction on bronchiolitis obliterans syndrome. Am J Respir Crit Care Med. 2007;175(5):507–13.
9. Huang HJ, et al. Late primary graft dysfunction after lung transplantation and bronchiolitis obliterans syndrome. Am J Transplant. 2008;8(11):2454–62.
10. Prekker ME, et al. Validation of the proposed international society for heart and lung transplantation grading system for primary graft dysfunction after lung transplantation. J Heart Lung Transplant. 2006;25(4):371–8.
11. Whitson BA, et al. Primary graft dysfunction and long-term pulmonary function after lung transplantation. J Heart Lung Transplant. 2007;26(10):1004–11.
12. Kreisel D, et al. Short- and long-term outcomes of 1000 adult lung transplant recipients at a single center. J Thorac Cardiovasc Surg. 2011;141(1):215–22.
13. Christie JD, et al. Construct validity of the definition of primary graft dysfunction after lung transplantation. J Heart Lung Transplant. 2010;29(11):1231–9.
14. Talaie T, et al. Ischemia-reperfusion injury in the transplanted lung: a literature review. Transplant Direct. 2021;7(2):e652.
15. de Perrot M, et al. Ischemia-reperfusion-induced lung injury. Am J Respir Crit Care Med. 2003;167(4):490–511.
16. Van Raemdonck D, et al. Report of the ISHLT working group on primary lung graft dysfunction Part IV: prevention and treatment: a 2016 consensus group statement of the International Society for Heart and Lung Transplantation. J Heart Lung Transplant. 2017;36(10):1121–36.

Fluids and Metabolics

14

Abstract

Volume assessment during lung transplantation is challenging due to altered physiology, surgical manipulation, and concurrent ECMO. Central venous pressure is unreliable, and dynamic measures such as pulse pressure variation are frequently confounded. Volume status must be inferred from a constellation of findings, including urine output, TEE assessment, lactate trends, and response to fluid boluses. Accurate assessment is essential to guide resuscitation, as both hypovolemia and fluid overload can lead to adverse outcomes, including primary graft dysfunction, organ hypoperfusion, and failure of extracorporeal support.

14.1 How Is Volume Status Estimated During Lung Transplant?

Estimating volume status during lung transplantation is challenging for several reasons, summarized in Table 14.1. Table 14.2 lists volume indicators which may be useful during lung transplant. As with all surgeries, volume assessment depends on a gestalt of these indicators.

M. Salvatori, A. Huang (eds.), *Anesthesia for Lung Transplant*,
https://doi.org/10.1007/978-3-032-06112-6_14

Table 14.1 Challenges in volume assessment in lung transplant

• Some patients have been aggressively diuresed preoperatively • Exposed pleural surfaces result in significant evaporative losses • Long and complex surgery associated with insensible fluid shifts • The CVP is routinely affected by surgical manipulation, preventing interpretation • Regular, abrupt changes in blood pressure can confound the interpretation of pulse pressure variation • Dampened or absent pulsatility on ECMO precludes arterial waveform analysis

Table 14.2 Indicators of volume status in lung transplant

• Urine output • TEE assessment of intracardiac volume • Hypotension responsive to fluid challenges • Trends in lactate, acid-base status, base excess • Trend in hemoglobin in the absence of significant surgical bleeding • ECMO “chugging”

14.2 What Is the Typical Volume Balance?

Double lung transplants routinely receive between 3 and 5 L of crystalloid/colloid during surgery. In general, transplants performed on ECMO receive 1-2 L more volume as temporary obstruction of venous return from surgical manipulation can prevent effective circulation by the extracorporeal circuit.

Few studies have examined the relationship between intraoperative fluid management and subsequent outcomes after lung transplantation. No studies have prospectively investigated fluid administration and inotropic/vasoactive agent use in perioperative period [1].

In a retrospective analysis of 118 lung transplants, Pilcher et al. (2005) showed that a high central venous pressure (CVP > 7 mmHg) following lung transplantation was associated with prolonged mechanical ventilation and higher ICU and hospital mortality. There were no complications associated with a CVP < 7 mmHg. The authors acknowledge that CVP is a poor indicator of intravascular volume status, as it is influenced by factors such as TR, RV dysfunction, and pulmonary pressures. The low CVP group (<7 mmHg) also had lower inotrope requirements and lower creatinine levels [2]. A retrospective cohort analysis of 494 lung transplants by Geube et al. (2017) showed that the median total volume administered was higher in patients who developed grade-3 PGD (5.0 L) than in patients with grades 0–2 PGD (3.8 L). Each intraoperative litre of fluid increased the odds of grade-3 PGD by approximately 22%. The use of vasoconstrictor and inotropic agents was greater in grade-3 PGD patients than in low-grade PGD patients. Type of fluid administered (crystalloids vs. colloids) did not influence the likelihood of PGD. Notably, this study included transplants performed with and without full CPB, but not ECMO. Increased volume administration was also associated with longer intensive care stay, but not time to extubation [3].

In contrast, in a retrospective cohort study of 1142 patients who underwent open thoracotomy and lung resection, Kim et al. (2020) showed that severe intraoperative fluid restriction and "zero-balance" fluid management was associated with an increased rate of complications including ARDS [4]. In their study, the fewest complications were recorded in patients who received an average of 4-5 ml/kg/hr for 24 hours from the start of surgery. The 2018 RELIEF trial (Restrictive versus Liberal Fluid Therapy for Major Abdominal Surgery) similarly showed that patients in the restrictive fluid group had a significantly higher risk of acute kidney injury than those in the liberal fluid group [5].

Overall, these findings are not yet robust enough to influence clinical practice. Furthermore, fluid management practices from other types of thoracic surgery (i.e., pneumonectomies and esophagectomies) cannot easily be translated to lung transplantation. As with all cases, a judicious and goal-directed fluid strategy is advised.

The most common approach to fluid repletion at TGH is a "moderately liberal" strategy, aiming for a positive fluid balance of 2-3 L at the end of surgery.

14.3 Are There Any Benefits to Albumin vs. Crystalloid?

The debate regarding crystalloid vs. colloid for volume replacement and resuscitation has spanned decades and divisions of medicine, and there remains little consensus or clinical evidence to direct practice. Only a single study has specifically investigated colloid administration during lung transplantation. McIlroy et al. (2009) reported an inverse relationship between colloid administration and early postoperative lung function. In their study of 107 transplants, administration of intraoperative colloid was associated with lower PF ratios 12 hours after surgery. The average volume of colloid administered in this study was 1.11 L. Each additional litre of intraoperative colloid was associated with a 35% reduction in the rate of extubation and a non-significant trend towards a reduced rate of ICU discharge [6]. Notably, the colloid administered in this study was Gelofusine, a 4% succinylated gelatin volume expander. It is not clear whether these findings translate to the use of albumin.

The common practice at TGH is to incorporate 5% albumin for ongoing fluid replacement after 2-3 L of crystalloid has been administered. This is not an evidence-based practice but is based on the theory that colloid solutions will increase oncotic pressure and thus remain in circulation for longer, and may also aid in preserving the endothelial glycocalyx leading to less pulmonary edema.

14.4 What Are the Risks of Volume Overload and Restriction? (Tables 14.3 and 14.4)

Table 14.3 Risks of volume overload in lung transplant

- Pulmonary edema, primary graft dysfunction, and hypoxemia
- Tissue edema compromising vascular and bronchial anastomoses
- RV dilatation and dysfunction
- Venous congestion of abdominal organs, particularly liver and kidneys
- Face and airway edema complicating tube exchange and post-op extubation
- Dilutional anemia and coagulopathy

Table 14.4 Risks of volume restriction in lung transplant

- Hypotension and high pressor requirements
- Low cardiac output
- Hypoperfusion and dysfunction of end organs, exacerbated by high pressors
- Lactic acidosis and biochemical derangements
- Poor ECMO circuit flow
- Acute kidney injury

14.5 Should Low Urine Output be Treated with Diuresis?

The significance of intraoperative oliguria has been investigated as a predictor of early AKI following major surgery. Although oliguria during lung transplantation has not been specifically studied, correlates can be drawn from other major surgeries. Given the high morbidity and mortality associated with dialysis after lung transplantation [7], early detection and prevention of renal injury is paramount.

In a single-centre retrospective observational study of 5894 patients undergoing major non-cardiac surgery, Shiba et al. (2018) found that intraoperative oliguria (UO < 0.5 mg/kg/hr for >120 minutes) was independently associated with postoperative AKI and increased serum creatinine [8]. The differences between groups were significant but modest, with 7.3% of oliguric patients progressing to AKI, versus 5.9% of non-oliguric patients. Cumulative duration of oliguria also increased the risk of developing postoperative AKI. Patient and operative characteristics associated with intraoperative oliguria included greater body weight, male sex, longer duration of anesthesia, and greater blood loss. Hydroxyethyl starch infusion ($P < 0.001$), use of furosemide ($P < 0.001$), postoperative renal replacement therapy requirement during hospital stay ($P < 0.001$), postoperative intensive care unit admission ($P = 0.001$), and incidence of hospital mortality ($P = 0.001$) were also more common in the ≥120-minute oliguria group. As this was a retrospective, observational study, it is not clear whether augmenting urine output with diuretics influences postoperative renal outcomes.

Similarly, in a retrospective analysis of 3560 patients undergoing major abdominal surgery, Mizota et al. [9] report an association between intraoperative oliguria and AKI, especially at thresholds <0.3 ml/kg/hr. Notably, there was no AKI association found if urine output was between 0.3–0.5 ml/kg/hr. These findings contrast with other retrospective studies on major abdominal surgery, which have shown that reduced intraoperative urine output does not predict postoperative AKI [10, 11]. To complicate the situation further, Kambhampati et al. [12] showed in a prospective study of 100 cardiac surgery patients that a positive fluid balance in the perioperative period was also associated with acute kidney injury and creatinine rise, with the highest quartile fluid balance group having a five-fold increased risk for AKI [12].

Specific to thoracic surgery, a randomized controlled trial of 102 patients undergoing elective VATS lung resection showed that intraoperative urine output was not associated with the amount of fluid administered across a range of 2-8 ml/kg/hr. The study also demonstrated that regardless of volume administered, reduced intraoperative urinary output (<1 ml/kg/hr) was not associated with postoperative renal dysfunction [13]. There were no differences in complications detected in the low vs. high volume groups.

Clearly, these discrepant results are difficult to apply clinically. As a result, there is marked practice variability among anesthesiologists regarding intraoperative diuresis to augment low urine output. A conservative approach would be to attempt low-dose diuresis (i.e., 10–20 mg of furosemide in a naïve patient) if urine output remains less than 0.3 ml/kg/hr. for >120min despite adequate volume resuscitation. However, be advised that the current evidence does not suggest that furosemide reduces the risk of renal replacement therapy or mortality in patients with acute kidney injury [14]. However, augmenting urine output may be useful in achieving fluid balance and reducing pulmonary edema in the critical peri-transplant period.

14.6 Should HCO_3 Be Administered to Correct Acid-Base Imbalance?

Bicarbonate (HCO_3) should not be used to manage respiratory acidosis, which can only be corrected by increasing minute ventilation. If the acidosis persists despite aggressive ventilation, consider ECMO support.

Bicarbonate should only be used to temporize severe metabolic acidosis while correcting the underlying cause. Common treatment triggers are pH $\leq$ 7.25, HCO3 $\leq$ 16, or BE < −5. HCO_3 dissociates into carbon dioxide and water, and 50 mEq (1amp) can transiently raise EtCO2 by as much as 20–30 mmHg. Although for most surgeries, this CO2 load is easily eliminated by increasing ventilation, the resultant hypercarbia can pose a significant challenge during lung transplant. Ventilatory compensation is difficult to achieve in the context of impaired gas exchange, increased dead space ventilation, and OLV, which may lead to persistent CO_2 accumulation and worsening respiratory acidosis. Increasing minute ventilation may be possible, but risks ventilator-induced lung injury. On ECMO, inform the perfusionist prior to administration to allow adjustments in sweep (typically increasing by 0.5 L/min). The differential for metabolic acidosis is broad, but rule out common causes such as hypovolemia, poor organ perfusion, hyperlactatemia (see Sect. 14.7), ketoacidosis, excessive saline administration (albumin 5% is reconstituted in normal saline), gastrointestinal (GI) losses, or patient-specific diagnoses (i.e., renal disease).

14.7 What Is the Differential and Management of Lactic Acidosis?

Lactate is a product of anaerobic metabolism, and climbing levels reflect inadequate oxygen delivery or reduced clearance by the liver. In non-ECMO transplants, lactate may rise due to persistent hypoxia and frequent periods of hemodynamic instability. Attempt to improve oxygenation by optimizing ventilation, adding iNO if necessary. Augment the blood pressure with fluid and vasopressors as indicated. Inform the surgeons as mechanical compression is a common contributor. If oxygenation and blood pressure remain inadequate, consider conversion to ECMO. For transplants on ECMO, climbing lactate more likely reflects hypoperfusion rather than hypoxia. Temporarily increase FiO2 to provide sufficient oxygenation during periods of reduced ECMO flow. Maintain a MAP between 60-80 mmHg and consider a trial of higher MAP (75-80 mmHg) if the lactatemia persists.

Iatrogenic causes of elevated lactate should also be examined. Suppression of insulin secretion by β2-agonists such as epinephrine, and increased hepatic insulin resistance after high-dose steroid administration leads to glycolysis and elevated pyruvate. Excessive pyruvate may overwhelm the Kreb's cycle even in the absence of anaerobic conditions, leading to increased lactate generation through lactate-pyruvate equilibrium [15–17].

Acute lung injury (ALI) is a recognized source of lactate production and should be suspected if lactate levels rise post-reperfusion and coincide with other signs of ALI, including hypoxia and pulmonary edema [18–20]. Mild ALI may be managed with optimization of ventilation (lung protective settings and individualized PEEP), iNO, and potentially diuresis. Severe ALI likely warrants discussion regarding ECMO support. Lactate levels that remain persistently elevated (> 5 mmol/L) despite adequate oxygenation and hemodynamics should prompt concern regarding extra-thoracic sources, such as gut ischemia from arterial emboli. Limb ischemia may occur due to peripheral ECMO cannulation, interstitial IV, or limb injury. These diagnoses are difficult to make and manage intraoperatively. Discuss your concerns with the surgeon and consider expedited transplantation, early specialist consultation (i.e., vascular, general surgery) and additional investigations (i.e., CT scan).

In lung transplantation, high lactate is associated with adverse early postoperative outcomes, including AKI, prolonged mechanical ventilation, and length of stay [18]. Elevated lactate at the end of surgery is also associated with grade 3 PGD on post-transplant day 3 [31]. Postoperative peak lactate levels within 72 hrs of transplantation are an independent predictor of 30-day and late mortality in lung transplant recipients [22].

14.8 What Is the Target Blood Sugar Range During Lung Transplant?

Perioperative hyperglycemia is consistently associated with adverse patient outcomes [23]. There are no evidence-based recommendations regarding intraoperative blood sugar management during lung transplantation, but correlates can be drawn from the cardiac surgery and ICU literature. The US Society of Thoracic Surgeons suggests a perioperative blood glucose level of <10 mmol/L for all cardiac surgery patients [24]. The American Diabetes Association recommends a target glucose range of 7.8–10 mmol/L for critically ill patients [25], whereas Diabetes Canada recommends a target of 6–10 mmol/L in critically ill patients [26].

The transplant literature shows that intraoperative hyperglycemia is associated with ischemia-reperfusion injury in kidney transplant [27], but not liver transplant [28]. However, intraoperative hyperglycemia during liver transplant is associated with increased postoperative infection and 1-year mortality. No studies have examined the impact of hyperglycemia on lung transplant outcomes. Given the current evidence, we recommend targeting a perioperative blood glucose level of 6–10 mmol/L for lung transplant.

14.9 What Is Hyperammonemia After Lung Transplant (HALT)?

Hyperammonemia is a rare but often fatal complication after lung transplant, with an incidence of 1–4% [29]. Thirty-day mortality in patients who develop HALT is 67%, compared to 17% for all other recipients [29]. The high mortality is partly due to delayed diagnosis, as the main symptom is altered level of consciousness, easily masked in patients intubated and sedated for several days for ventilatory reasons. Ammonia levels are therefore drawn for all lung transplant recipients as part of postoperative ICU care. Lung transplants are not exclusively affected, as hyperammonemia also occurs after orthotopic liver transplants and bariatric surgery [29]. The etiology remains unclear, but the dominant hypothesis is that opportunistic urea and ammonia-producing organisms, such as *Mycoplasma* and *Ureaplasa*, thrive in immunosuppressed patients, becoming a significant source of exogenous production [30, 31]. An alternative theory suggests that the stress of surgery unmasks enzymatic deficiencies in the urea cycle and glutamine synthetase pathways, exacerbated by calcineurin inhibitors and steroids [32, 33].

Regardless of etiology, high levels of ammonia enter the CNS causing cerebral edema, raised ICP, and encephalopathy. Acute hyperammonemia has more profound effects than the chronic form seen in end-stage liver failure, due to the absence of compensatory metabolic adaptations [34]. If not identified and treated promptly, long-term sequelae include refractory status epilepticus, long-term motor and cognitive impairment, and death [29]. The tenants of treatment are neuroprotection, anti-inflammatories, antibiotics, hemodialysis, excretion, and arrest of ammonia production [35]. Hemodialysis is critically important, as a 2017 systemic review of HALT reported a mortality of 100% without dialysis, 75% with continuous venovenous hemodialysis (slower rate of removal), and 40% with intermittent hemodialysis (faster rate of removal) [36]. Although laxatives like lactulose are established treatments for hyperammonia from liver failure, they do not reduce mortality in HALT [34]. All recipients with hyperammonemia receive a 2-week course of moxifloxacin and doxycycline without major modifications in the anti-rejection regimen.

14.10 Does Calcium Cause Ischemia-Reperfusion Injury?

Theoretically, the intermittent hypoxia during lung transplantation causes depletion of mitochondrial ATP. As a result, anaerobic cellular metabolism is initiated and decreased ATP impairs normal functioning of the Na/K pump, resulting in calcium sequestration inside the cell [37]. The intracellular calcium overload can lead to generation of reactive oxygen species (ROS), leading to ischemia-perfusion injury [37]. However, no clinical studies have shown that calcium administration worsens ischemic reperfusion injury during organ transplantation.

14.11 Should Calcium Be Replaced on ECMO?

In cardiac surgery, calcium administration is avoided on full CPB as it may promote coagulation, myocardial tone, and energy expenditure during hypothermic diastolic cardiac arrest. These considerations do not apply to lung transplantation on or off ECMO, and so calcium should be replaced whenever depleted.

References

1. Colombo J, et al. Lung transplantation is much more than lungs and fluids. Anesth Analg. 2016;123(3):794–5.
2. Pilcher DV, et al. High central venous pressure is associated with prolonged mechanical ventilation and increased mortality after lung transplantation. J Thorac Cardiovasc Surg. 2005;129(4):912–8.
3. Geube MA, et al. Increased intraoperative fluid administration is associated with severe primary graft dysfunction after lung transplantation. Anesth Analg. 2016;122(4):1081–8.
4. Kim JA, et al. Restrictive intraoperative fluid management was associated with higher incidence of composite complications compared to less restrictive strategies in open thoracotomy: a retrospective cohort study. Sci Rep. 2020;10(1):8449.
5. Myles PS, et al. Restrictive versus Liberal fluid therapy for major abdominal surgery. N Engl J Med. 2018;378(24):2263–74.
6. McIlroy DR, et al. Does anaesthetic management affect early outcomes after lung transplant? An exploratory analysis. Br J Anaesth. 2009;102(4):506–14.

7. Mason DP, et al. Dialysis after lung transplantation: prevalence, risk factors and outcome. J Heart Lung Transplant. 2007;26(11):1155–62.
8. Shiba A, et al. Association between intraoperative oliguria and acute kidney injury after major noncardiac surgery. Anesth Analg. 2018;127(5):1229–35.
9. Mizota T, Yamamoto Y, Hamada M, Matsukawa S, Shimizu S, Kai S. Intraoperative oliguria predicts acute kidneyinjury after major abdominal surgery. Br J Anaesth. 2017;119(6):1127–34. https://doi.org/10.1093/bja/aex255. PMID:29136086
10. O'Connor ME, et al. Incidence and associations of acute kidney injury after major abdominal surgery. Intensive Care Med. 2016;42(4):521–30.
11. Grams ME, et al. Acute kidney injury after major surgery: a retrospective analysis of veterans health administration data. Am J Kidney Dis. 2016;67(6):872–80.
12. Kambhampati G, et al. Perioperative fluid balance and acute kidney injury. Clin Exp Nephrol. 2012;16(5):730–8.
13. Matot I, et al. Fluid management during video-assisted thoracoscopic surgery for lung resection: a randomized, controlled trial of effects on urinary output and postoperative renal function. J Thorac Cardiovasc Surg. 2013;146(2):461–6.
14. Ho KM, Power BM. Benefits and risks of furosemide in acute kidney injury. Anaesthesia. 2010;65(3):283–93.
15. Bakker J, et al. Lactate-guided resuscitation saves lives: we are not sure. Intensive Care Med. 2016;42(3):472–4.
16. Bakker J, et al. Lactate: where are we now? Crit Care Clin. 2020;36(1):115–24.
17. Kraut JA, Madias NE. Lactic acidosis. N Engl J Med. 2014;371(24):2309–19.
18. Worrell SG, et al. Is lactic acidosis after lung transplantation associated with worse outcomes? Ann Thorac Surg. 2020;110(2):434–40.
19. Kellum JA, et al. Release of lactate by the lung in acute lung injury. Chest. 1997;111(5):1301–5.
20. De Backer D, et al. Lactate production by the lungs in acute lung injury. Am J Respir Crit Care Med. 1997;156(4 Pt 1):1099–104.
21. Fessler J, et al. Blood lactate level evolution during lung transplantation. J Heart Lung Transplant. 2021;40(4):S317.
22. Xu J, et al. The prognostic value of peak arterial lactate levels within 72 h of lung transplantation in identifying patient outcome. J Thorac Dis. 2020;12(12):7365–73.
23. Duggan EW, et al. Perioperative hyperglycemia management: an update. Anesthesiology. 2017;126(3):547–60.
24. Lazar HL, et al. The Society of Thoracic Surgeons practice guideline series: blood glucose management during adult cardiac surgery. Ann Thorac Surg. 2009;87(2):663–9.
25. American DA. Standards of medical Care in Diabetes-2021 abridged for primary care providers. Clin Diabetes. 2021;39(1):14–43.
26. Diabetes Canada Clinical Practice Guidelines Expert C, et al. In-hospital management of diabetes. Can J Diabetes. 2018;42(Suppl 1):S115–S23.
27. Parekh J, et al. Intraoperative hyperglycemia augments ischemia reperfusion injury in renal transplantation: a prospective study. J Transp Secur. 2011;2011:652458.
28. Kumar SS, et al. Intraoperative glycemic control in patients undergoing Orthotopic liver transplant: a single center prospective randomized study. BMC Anesthesiol. 2020;20(1):3.
29. Leger RF, et al. Hyperammonemia post lung transplantation: a review. Clin Med Insights Circ Respir Pulm Med. 2020;14:1179548420966234.
30. Bharat A, et al. Disseminated Ureaplasma infection as a cause of fatal hyperammonemia in humans. Sci Transl Med. 2015;7(284):284re3.
31. Wylam ME, et al. Fatal hyperammonaemia caused by mycoplasma hominis. Lancet. 2013;382(9908):1956.
32. Castineira J, et al. Postbariatric surgery Hyperammonemia: a rare cause of encephalopathy. ACG Case Rep J. 2019;6(7):e00119.
33. Seethapathy H, Fenves AZ. Pathophysiology and Management of Hyperammonemia in organ transplant patients. Am J Kidney Dis. 2019;74(3):390–8.
34. Clay AS, Hainline BE. Hyperammonemia in the ICU. Chest. 2007;132(4):1368–78.

35. Said VJ, Garcia-Trujillo E. Beyond lactulose: treatment options for hepatic encephalopathy. Gastroenterol Nurs. 2019;42(3):277–85.
36. Krutsinger D, et al. Idiopathic hyperammonemia after solid organ transplantation: primarily a lung problem? A single-center experience and systematic review. Clin Transpl 2017;31(5).
37. Fernandez AR, et al. Review: ischemia reperfusion injury-a translational perspective in organ transplantation. Int J Mol Sci 2020;21(22).

Coagulation and Transfusion

15

Abstract

Blood transfusion is surprisingly common in the peri-transplant period, although rates have declined with the shift from cardiopulmonary bypass to ECMO. At TGH, two-thirds of patients receive pRBCs intraoperatively or within 24 h. Tranexamic acid appears safe in thoracic and transplant surgery, with evidence supporting early administration (≤50 mg/kg) to reduce blood loss. Transfusion decisions are individualized, guided by hemodynamics, bleeding, and coagulation targets.

15.1 How Common Is Transfusion During Lung Transplant?

Blood transfusion during lung transplantation has lessened with the transition from full CPB to ECMO. The documented rates in the literature vary by institution and local practices, but Pena et al. report the majority of recipients receive at least 3 units of PRBCs in the perioperative period [1]. At TGH in 2020, 67% of lung transplant recipients required pRBC transfusion intraoperatively or within 24 h of surgery, with a median of 2 pRBC per patient. The transfusion rate has increased yearly, from 58% in 2018 and 64% in 2019. Increased use of intraoperative ECMO support may contribute.

M. Salvatori, A. Huang (eds.), *Anesthesia for Lung Transplant*,
https://doi.org/10.1007/978-3-032-06112-6_15

15.2 Should Tranexamic Acid (TXA) Be Used for Lung Transplantation?

There are no systematic RCTs investigating whether TXA is safe to administer during lung transplantation, or whether it impacts transfusion rates [1], but corollary evidence can be drawn from the surgical, cardiothoracic, transplant, and ECLS literature.

The ideal TXA dosing strategy remains controversial, despite numerous randomized controlled trials and observational studies. In one of the largest studies of TXA to date, the POISE-3 trial demonstrated that a 1 g intravenous bolus of TXA at the beginning and end of surgery significantly reduced bleeding in non-cardiac surgery, although the effects were modest and noninferiority was not established for the primary safety outcome, composite cardiovascular events [2].

The OPTIMAL randomized control trial showed that for patients undergoing cardiac surgery, a high TXA dose regimen of 30 mg/kg bolus and 16 mg/kg/hr. infusion modestly reduced RBC transfusion rates compared to a low dose regimen of 10 mg/kg and 2 mg/kg/hr infusion [3]. There were no statistically significant differences in adverse events including 30-day mortality, seizure, kidney dysfunction, or thrombotic events. Even studies using single bolus doses of 50 to 100 mg/kg did not show differences in rates of death, stroke, pulmonary embolism, renal failure, or bowel infarction, although the likelihood of seizures increased [4]. Small, single-center studies have also shown TXA to be safe and effective in elective thoracic surgery [5].

Early studies of TXA in orthotopic liver transplant showed that high-dose TXA up to a maximum dose of 20 g reduced intraoperative blood loss, platelet, and cryoprecipitate requirement without clinical evidence of hepatic artery or portal vein thrombosis [6]. A more recent observational study in orthotopic liver transplantation showed that a tranexamic dose of over 10 mg/hr for more than half the duration of surgery results in reduced rates of RBC and FFP transfusions without any associated increase in thromboembolic events [7].

The ECLS literature describes safe use of TXA for ECMO patients, although this is typically used in the context of acute hemorrhage [8, 9]. Dose-finding studies have not been published in this population.

Taking all this evidence into consideration, it seems safe and effective to administer up to 50 mg/kg of tranexamic acid during major surgery, either via infusion or intermittent boluses. TXA is most effective when given early in the surgery (i.e., before skin incision), to preserve fibrinogen before lysis from surgical stresses and contact activation from the ECMO circuit.

15.3 What Are the Common Transfusion Triggers?

The decision to transfuse products depends on several factors including the rate and amount of ongoing bleeding, the presumed cause (surgical vs. coagulopathy), hemodynamic instability, remaining duration of surgery, oxygen demand, and comorbidities intolerant of anemia (i.e., severe coronary artery disease). As a rough guide, consider targeted transfusion of blood products if Hb < 80 g/L, PLT < 80, fibrinogen <1.5, and INR > 1.7. Remember that TRALI is a potential risk factor for PGD, and deranged coagulation values can often be treated expectantly in the absence of clinically relevant bleeding.

15.4 Can Blood Products Be Safely Administered on ECMO?

The Extracorporeal Life Support Organization (ELSO) guidelines specifically endorse the use of pRBCs, platelets, antifibrinolytics, and fresh frozen plasma for patients with continued bleeding on ECMO [10]. Octaplex (or other Prothrombin Complex Concentrates) is rarely required and should be discussed with the surgical team prior to administration.

15.5 Is There Benefit to Octaplex vs. Fibrinogen vs. FFP?

The specific blood product you choose to administer will depend on laboratory values, Rotational Thromboelastometry (ROTEM), and volume tolerance. Octaplex replaces vitamin K-dependent factors and corrects INR with minimal volume administration. However, if the clotting time remains prolonged, appreciate that other factors may be deficient, warranting FFP. Similarly, if only fibrinogen is deficient, 4 g of Fibrinogen Concentrate contains the same amount of fibrinogen as 4 units of FFP or 10 units of cryoprecipitate, but with one-quarter of the volume administration.

15.6 When Should ROTEM Be Considered?

Although not routine during lung transplantation, request a ROTEM for any of the following scenarios:

1. Persistent bleeding without an identifiable surgical source
2. Markedly deranged coagulation values on laboratory testing
3. Known risk factors for coagulopathy or platelet dysfunction (recent blood thinners, antiplatelet agents, end stage renal failure (ESRF))
4. Known pro-thrombotic risk factors (antiphospholipid antibody (APLA), COVID)
5. Massive transfusion protocol (i.e., inadvertent vascular or cardiac injury, redo transplant)

15.7 Should Heparin Be Given via Bolus or Infusion for ECMO?

Intermittent heparin boluses vs. continuous heparin infusion for maintenance of anticoagulation on ECMO depends on surgeon preference. Most surgeons prefer bolus administration unless they anticipate prolonged surgical time or continued ECMO post-op. Initial bolus dose is typically 50–100 u/kg. Standard intermittent bolus doses and infusion rates are both 10–20 u/kg/hr. Target an ACT >220 for cannulation, then maintain ACT 160–180 for duration of ECMO.

15.8 When Is Protamine Administered?

Heparin reversal with protamine is typically not required following separation from ECMO, and the ACT is allowed to normalize over time. Discuss protamine together with surgeons if there is persistent bleeding or coagulopathy, or an ACT >250.

References

1. Pena JJ, et al. Perioperative Management of Bleeding and Transfusion for lung transplantation. Semin Cardiothorac Vasc Anesth. 2020;24(1):74–83.
2. Devereaux PJ, et al. Tranexamic acid in patients undergoing noncardiac surgery. Reply. N Engl J Med. 2022;387(9):858–9.
3. Shi J, et al. Effect of high- vs low-dose Tranexamic acid infusion on need for red blood cell transfusion and adverse events in patients undergoing cardiac surgery: the OPTIMAL randomized clinical trial. JAMA. 2022;328(4):336–47.
4. Myles PS, et al. Tranexamic acid in patients undergoing coronary-artery surgery. N Engl J Med. 2017;376(19):1893.
5. Casati V, et al. Hemostatic effects of tranexamic acid in elective thoracic aortic surgery: a prospective, randomized, double-blind, placebo-controlled study. J Thorac Cardiovasc Surg. 2002;123(6):1084–91.
6. Boylan JF, et al. Tranexamic acid reduces blood loss, transfusion requirements, and coagulation factor use in primary orthotopic liver transplantation. Anesthesiology. 1996;85(5):1043–8; discussion 30A-31A.
7. Badenoch A, et al. The effectiveness and safety of Tranexamic acid in Orthotopic liver transplantation clinical practice: a propensity score matched cohort study. Transplantation. 2017;101(7):1658–65.
8. Lotz C, et al. Therapeutic interventions and risk factors of bleeding during extracorporeal membrane oxygenation. ASAIO J. 2017;63(5):624–30.
9. Ratano D, Fan E. Thrombotic Events in Extracorporeal Life Support (ECLS) Patients Treated with Tranexamic Acid (TXA) for Bleeding: a retrospective study. TP56 TP056 Nonpulmonary Critic Care Lung Infect. 2021:A2876–A76.
10. Organization ELS. General Guidelines for all ECLS Cases 2017; (Version 1.4). Accessed 20 Nov 2022.

Transfer to ICU

16

Abstract

Transfer to ICU requires careful planning and encompasses tube exchange, ETT clamping during circuit transitions, line security, transport medications, and ECMO pump coordination. A standardized checklist ensures airway integrity, medication continuity, and documentation. Thorough ICU handover should cover graft, anesthetic, surgical, and hemodynamic details to ensure safe continuity of care.

16.1 When Should the DLT Be Changed for an SLT?

At the end of the transplant, exchanging the DLT to an SLT allows for better bronchoscopy and suctioning, minimizes the risk of tracheal injury and subglottic stenosis, and facilitates postoperative ICU care. The exchange is a high-risk procedure that can lead to hypoxia, hypercarbia, hemodynamic instability, and cardiac arrest if not well planned and executed. This procedure should be undertaken during a defined period, where no other activities are taking place (surgical maneuvers, room turnover, etc.). Adequate personnel (staff anesthesiologist and fellow, staff surgeon) and equipment should be available and prepared ahead of time (Table 16.1).

For lung transplants performed off ECMO, the DLETT is exchanged for a single lumen tube at the end of the procedure once the drapes have been removed and the patient can be optimally positioned. In extremely rare cases, when the tube exchange seems too perilous to perform safely due to airway trauma or gross edema, you may discuss transfer of the patient to ICU with the DLETT in situ. These cases require thorough documentation, communication, and a clear plan for eventual tube exchange (i.e., following diuresis or steroid therapy for airway inflammation, with anesthesia support).

M. Salvatori, A. Huang (eds.), *Anesthesia for Lung Transplant*,
https://doi.org/10.1007/978-3-032-06112-6_16

Table 16.1 Algorithm for DLETT to SLT tube exchange

- Call for an experienced assistant (second staff, anesthesia fellow).
- Ensure a proper sized SLT, face mask, oral airway, tube exchange catheter, and bronchoscopy adapter are readily available.
- Check that the adult bronchoscope has been connected to the monitor with suction connected.
- Consider readministration of muscle relaxant if not done recently.
- **OFF ECMO:** Increase FiO_2 to 100% for 2–3 min prior to exchange.
- **ON ECMO:** Request the perfusionist increase flows to full and O_2 to 100% for duration of tube exchange.
- Suction out oropharynx.
- Deflate the endobronchial cuff if not already done.
- Insert CMAC video laryngoscope into the oropharynx and visualize glottis. Inspect for significant swelling. Suction any remaining secretions. Keep CMAC blade in place during exchange. **Do not proceed if there is significant swelling, or if glottic visualization is poor.**
- Stop ventilator (set to manual, or use pause gas flow function).
- Insert a tube exchange catheter through the *tracheal* lumen of the DLETT, advancing until the depth markers on the catheter read 3-5 cm greater than the length of the DLETT.
- Deflate tracheal cuff.
- In coordination with assistant, slowly remove DLETT while advancing the tube exchange catheter at the same pace. Observe with videolaryngoscope.
- Once the DLETT has been removed, advance the SLT over the tube exchange catheter to a depth of 20-23 cm. Observe passage of SLT through the cords with the CMAC.
- Inflate SLT cuff and resume ventilation. Confirm $ETCO_2$.
- Have surgical team do bronchoscopy to inspect anastomoses, confirm final ETT position, and to perform pulmonary toileting.
- Secure ETT with new tape or ties.
- Perform a recruitment maneuver and resume ventilation.
- **ON ECMO:** Inform perfusionist that ECMO flows can be restored to previous.

For lung transplants on ECMO, the timing of tube exchange depends on staff preference, although it is generally safer to exchange the tube on ECMO to avoid crisis should the exchange prove difficult. Additional benefits include easier and more effective bronchopulmonary toilet through the SLT prior to discontinuation of ECMO. Limitations of this approach include awkward access under the drapes, poor positioning, and the inability to subsequently isolate lungs during chest closure. Conversely, waiting until the end of the case allows for lung isolation during closure, unobstructed ETT access, and optimal positioning, but at the risk of life-threatening desaturation should the exchange fail.

Be mindful that the use of TEE will impact the timing and difficulty of tube exchange. For example, exchanging the tube on ECMO will typically require removal of the transesophageal echo probe, precluding a full echocardiographic assessment once the patient is off ECMO. Removal and reinsertion of the TEE probe multiple times during the case is typically avoided due to the increased risk of oropharyngeal or esophageal injury.

Following the exchange, ensure the airway is properly secured and the ETT depth is appropriate. End-of-case bronchoscopy is routinely performed by the surgical team to inspect the anastomoses and perform any necessary toileting. This provides an opportunity to confirm ETT depth.

16.2 How Are Patients Prepared for Transfer to ICU?

Lung transplant recipients are transferred to ICU for post-transplant monitoring, weaning from mechanical ventilation and ongoing care. The transfer to ICU can take 10–15 min, and thorough preparation and planning is required.

Airway Ventilation during transfer is via self-inflating Ambu bag with a PEEP valve or portable ventilator. Always bring a facemask in the rare event of unexpected extubation during transfer. Despite the emphasis on minimizing oxygen exposure during the transplant, all patients are transferred on 100% FiO2. When changing between mechanical and manual ventilation, clamp the ETT prior to disconnection to prevent derecruitment. This step is particularly important in patients with high FiO2 requirements, early PGD, or large body habitus. An RT/AA is required for all transports involving iNO.

Lines Mount all infusion pumps, fluid bags, and pressure transducers onto a single IV pole for transfer. All lines should be cleanly dressed. Central and PA lines should be secured to the patient's chest/shoulder with adhesive tape to reduce the risk of inadvertent migration or removal. Ensure line length is sufficient to allow positioning on the right-hand side of the patient in the ICU. Confirm that the PA catheter balloon is deflated, and the balloon inflation port is locked. Ensure that fluid is infusing through the side-arm of the Cordis during transport as the ICU may not use the side-arm if clamped for over 20 min. The surgeons manage the chest tubes and ensure proper placement of pleurovacs and Foley for transfer. Should the patient require ECMO post-op, the perfusionists secure the cannulas and pump and accompany the transport.

Medications Patients are transported with the infusions used throughout the surgery, which typically include propofol, norepinephrine, and vasopressin. Ensure that there is enough medication for at least 60 min to cover the duration of transport, handover, and ICU admission. Discontinue sufentanil and MgSO4 infusions as these are not standard in the ICU. Transport medications should also include emergency syringes of propofol, epinephrine, and phenylephrine.

16.3 What Is the Pre-transport Checklist Prior to Leaving the OR? (Table 16.2)

Table 16.2 Pre-transport checklist

System	Checklist items
Airway	DLT exchanged for SLT Bronchoscopy by surgical team—Check ETT position, fix tube, note depth Facemask, PEEP valve, portable suction available
Breathing	Full O2 cylinder Ambu bag ETT clamp for circuit changes Call RT if on iNO
Circulation	Redress soiled lines Secure PA and central line to patient's shoulder All transducers, pumps, and IV bags to a single pole Confirm sufficient infusion volume to last 60 min Full fluid bags Fluid infusing through side arm of Cordis Perfusionist secures lines and pump if patient on ECMO

Others	Insert nasogastric tube
	Complete all documentations including:
	  LAS variables (emailed)
	  Blood transfusions
	  Fluid balance
	  Medications administered
	  Procedure notes
	  Pain Service consult
	Emergency drugs:
	  20 cc syringe of phenylephrine (40 ug/cc)
	  10 cc syringe of epinephrine (4 ug/cc)
	  20 cc of propofol (10 mg/cc)

16.4 What Information Is Included in the ICU Handover? (Table 16.3)

Table 16.3 ICU handover checklist

Category	Checklist items
Patient background	Patient age Indication for transplant and baseline functional status Presence of pulmonary hypertension Any other significant comorbidities
Graft details	Single or double lung transplant Donor type EVLP requirement
Anesthetic details	Events on induction Intubation grade and equipment used Lines, placement, need for CXR confirmation
Surgical details	ECLS use and weaning Indication for post-transplant ECLS if applicable Total heparin dose, final ACT, protamine requirement Pulmonary artery pressures post-transplant Fluid balance Intraoperative pressor requirements Blood loss and transfusion requirement Any complications (i.e., PGD, arrhythmias)
Intraoperative investigations	TEE findings ROTEM and coagulation studies Final ABG values
Ongoing management	Final vent settings, including PEEP and FiO2 Current pressor requirements Last dose of antibiotics Last muscle relaxant Total opioids, regional anesthesia, pain issues Any issues that could complicate extubation, i.e., difficult tube exchange, airway swelling, or edema Pending or incomplete investigations Any outstanding management details Suitability for vent and sedation weaning

17 Special Situations

Abstract

Lung transplantation presents unique anesthetic and surgical challenges across complex patient populations. In cystic fibrosis, meticulous airway management and infection control are critical due to copious secretions and multidrug-resistant organisms. Single lung transplantation requires nuanced ventilation strategies to avoid native lung hyperinflation or injury to the graft. Bridging with V-V ECMO demands careful anticoagulation and hemodynamic vigilance. Lung transplant for COVID-ARDS hinges on candidate selection and timing. Combined heart-lung transplants require coordinated, multi-specialty approaches throughout cardiopulmonary bypass. Redo transplantation carries higher risks due to adhesions, coagulopathy, and prolonged operative time.

17.1 Cystic Fibrosis

The TLTP is the largest and most experienced cystic fibrosis lung transplant program, having performed the first CF lung transplant in 1986. Cystic fibrosis presents several unique challenges from an anesthetic management perspective, including patient age and the multi-systemic/extra-pulmonary manifestations of CF (Table 17.1). The infectious/suppurative nature of CF necessitates a double lung transplant.

Survival after lung transplantation in CF patients is heavily linked to factors such as *Burkholderia cepacia complex (BCC)* infection, younger age (poorer outcome), and pre-transplant functional status [1]. *BCC* is a family of greater than 20

M. Salvatori, A. Huang (eds.), *Anesthesia for Lung Transplant*,
https://doi.org/10.1007/978-3-032-06112-6_17

Table 17.1 Anesthetic considerations for cystic fibrosis

System	Manifestation
Upper airway	• Sinusitis • Nasal polyps
Pulmonary	• Goblet cell hypertrophy + decreased mucociliary clearance ⇒ atelectasis, hypoxia, airway inflammation • Bacterial colonization: *Pseudomonas, S. aureus, H. influenza, Stenotrophomonas, Burkholderia cepacia complex,* and aspergillus • Recurrent/chronic infection ⇒ bronchiectasis + hemoptysis + fibrosis • Mixed restrictive/obstructive lung physiology • Cystic lung disease ⇒ pneumothorax • Chronic hypoxia ⇒ increased PVR/pulmonary HTN +/− right heart failure
Gastrointestinal	• Gastroesophageal reflux disease (GERD) • Distal intestinal obstruction syndrome (DIOS)
Endocrine	• Exocrine dysfunction ⇒ malabsorption + emaciation + fat soluble vitamin deficiency (including vitamin K ⇒ potential for coagulopathy) • Exocrine dysfunction ⇒ pancreatitis • Endocrine dysfunction ⇒ diabetes and microvascular complications
Hepatobiliary	• Fatty infiltration ⇒ cirrhosis • Cholelithiasis/cholecystitis
MSK	• Hypertrophic osteoarthropathy • Osteoporosis • Arthritis
Other	• Small stature • Reproductive difficulties • Antibiotic resistance and sensitization (allergies)

gram-negative bacteria known to infect/colonize the immunocompromised. In CF patients, *BCC* infection causes rapid deterioration in lung function, and can produce "cepecia syndrome" consisting of necrotizing pneumonia, respiratory failure, bacteremia, and death over a course of days [2]. The TLTP experience in transplanting CF patients has shown 1-, 5-, and 10-year survival rates of 94%, 70%, and 53% respectively in *BCC*-negative patients. Conversely, *BCC*-infected recipients had significantly poorer survival rates (1-, 5-, and 10-year survival of 59%, 33%, and 16%, respectively) [1]. Recall that 5-year survival for all patients having undergone lung transplantation is approximately 50%. *BCC* is exceedingly difficult to eradicate.

Pre-induction

Many CF patients will present with a Port-a-Cath or Peripherally Inserted Central Catheter (PICC) line used for chronic antibiotic therapy. The decision to keep the pre-existing PICC depends on time since insertion, complications (blocked ports, concerns regarding infection), and ease of central venous access overall. Discuss with the transplant team.

Additional Equipment
Single-lumen ETT, adult bronchoscope, ample irrigation, poviodine.

Induction
As per usual, bearing in mind the relative younger age and smaller stature of CF patients relative to the typical transplant recipient.

Airway
Due to the suppurative nature of CF, copious secretions are present in the airways, even with regular chest physiotherapy and humidified treatments. As such, intubation is typically done using a single-lumen tube at first, to allow for a thorough pulmonary toileting prior to insertion of a double-lumen tube for the transplant procedure. Consider exchange to DLT after line insertion (see below).

Position
Supine, arms out.

Incision
Clamshell.

Immunosuppression
500 mg Solumedrol at induction.

Antibiotic
The broad range of colonizing organisms, combined with the frequent occurrence of antibiotic sensitization, produces significant variation in antibiotic regimens. Overall, dual *Pseudomonas* coverage is indicated, and Ceftazidime and Meropenem are commonly used antibiotics.

Disposition
ICU.

Operative Course

Following induction and intubation with a single lumen tube, the surgical team will perform a thorough pulmonary toilet. Although the SLT is often exchanged for a DLT at this point, consider leaving the SLT for line insertion. In the 30–40 minutes required for lines, the patient will be in a head-down position and likely express more purulent secretions that may obstruct the smaller lumens of a DLT. Leaving the SLT in place minimizes the risk of obstruction and allows for further toileting prior to DLT exchange.

The CF patient poses significant anesthetic and surgical challenges. The mixed restrictive and obstructive pathophysiologies, worsened by the need for OLV and the increased dead space associated with positive pressure ventilation, result in an increased propensity for hypoxia and hypercarbia/acidosis. At the same time, the relatively small hemithoraces and potential presence of adhesions from prior procedures/chest tubes make dissection difficult and prone to hemodynamic challenges from surgical retraction. As such, VA ECMO use during CF transplants is common.

Ventilation and OLV management of the native lung(s) may require higher airway pressures and respiratory rates to achieve acceptable oxygenation and ventilation. Permissive hypercapnia is often necessary despite aggressive ventilator settings. Progressive hypoxia or hypercarbia should favor early ECMO support.

Because CF patients are colonized with multi-drug-resistant bacteria, absolute lung isolation with a well-placed DLT is required. Maintenance of strict isolation is necessary to avoid any contamination of the new implant, and manipulation of the DLT should be done with care.

Additionally, the TLTP protocol involves irrigation of the airway and pleural spaces with an iodine solution following pneumonectomy and prior to implantation of the graft. There is no strong evidence supporting long-term benefits of this practice, but the potential for reduced shorter-term infections and the relatively low risk associated with the process favors its application [3]. 50cc of iodine prep solution is diluted in a total volume of 250-300cc normal saline. 50cc of the diluted iodine solution is instilled down each lumen of the DLT, followed by 50cc of normal saline, flushed and suctioned by the surgical team.

Bleeding may be more significant in CF transplants due to intrinsic coagulopathy (malnutrition and vitamin malabsorption), adhesions from prior procedures, and the dilution and anticoagulation associated with ECMO. Perform frequent CBC, coagulation panels, and ROTEM as indicated.

Due to the potential presence of nasal polyps (up to 48% of CF patients) [4] and potential for nasal obstruction due to a history of rhinosinusitis (up to 80% of CF patients) [4], placement of the nasogastric tube can be challenging and traumatic. In patients who have been on ECMO and present with a history of known or suspected nasal polyps or nasal obstruction, nasogastric tubes should be avoided due to the risk for significant epistaxis. An orogastric tube is a safe and reasonable alternative.

17.2 Single Lung Transplant

In general, single lung transplantation (SLT) is reserved for patients with pulmonary fibrosis over the age of 65, as they show equivalent survival to DLTs [5]. SLT for COPD varies by center, but the TLTP favors DLT for COPD, owing to possible survival benefit [6]. Chronic infection and colonization preclude patients with cystic fibrosis and bronchiectasis from SLT. Candidates with primary pulmonary arterial hypertension also require DLT due to bilateral involvement of the pulmonary vasculature.

Additional Equipment
None.

Pre-induction, Induction
As per double lung transplant.

Airway
Intubation with a double-lumen tube. For a left-sided single lung transplant, a right-sided DLETT minimizes interference with the left-sided bronchial anastomosis. However, right-sided DLETTs can be challenging and positional, and ineffective if the origin of the right upper lobe is within 1 cm of the carina. In case of difficulty, a left-sided single lung transplant can be safely and easily accomplished with a left-sided DLETT with some minor repositioning, as in double lung transplants.

Lines
Consider subclavian access on the surgical side to minimize risk of injury to the remaining native lung.

Position and Incision
Single lung transplant can be performed in the supine position with a hemiclamshell incision, or in the lateral decubitus position via a posterolateral thoracotomy incision.

The supine approach with hemiclamshell incision is more commonly used and involves an anterior thoracotomy and transverse sternotomy with

(continued)

continued extension partway across the contralateral hemithorax. The hemi-clamshell provides adequate exposure for the transplant and central ECMO cannulation if needed. The transverse sternotomy can be avoided in patients with larger hemithoraces (COPD), and during transplants where the use of ECMO is not considered likely.

The lateral approach is less common, but more advantageous in patients who have had prior thoracic surgery. Lateral positioning also optimizes V/Q matching during OLV but precludes central VA-ECMO. Peripheral cannulation can be obtained via the ipsilateral groin.

Immunosuppression
500 mg Solumedrol at induction.

Antibiotic
Piperacillin/Tazobactam (4.5 g) q6h.

Disposition
ICU.

Operative Course
Ventilation during a single lung transplant can present a unique challenge. Prior to reperfusion, oxygenation and ventilation will depend on the remaining native lung. As this lung will remain in situ, care should be taken to avoid the aggressive ventilation of native lungs common in double-lung transplants. Lung protective ventilation (4-5 cc/kg), permissive hypercapnia, and titration of FiO_2 to an SpO_2 of ~92–94% minimizes the risk of acute lung injury in the native lung. If adequate oxygenation or ventilation is not sustainable with lung protective settings, conversion to ECMO may be required.

Following reperfusion, two lung ventilation is complicated by drastic differences in compliance between lungs, evidenced by biphasic capnography (Fig. 17.1). In the setting of COPD, the native lung is typically more compliant and receives preferential ventilation. This "Native Lung Hyperinflation" has been reported to occur in up to 30% of single lung transplants for COPD [7]. Native lung hyperinflation can cause significant mediastinal shift and hemodynamic instability, as well as compression and atelectasis of the transplanted lung. Intraoperatively, this can be

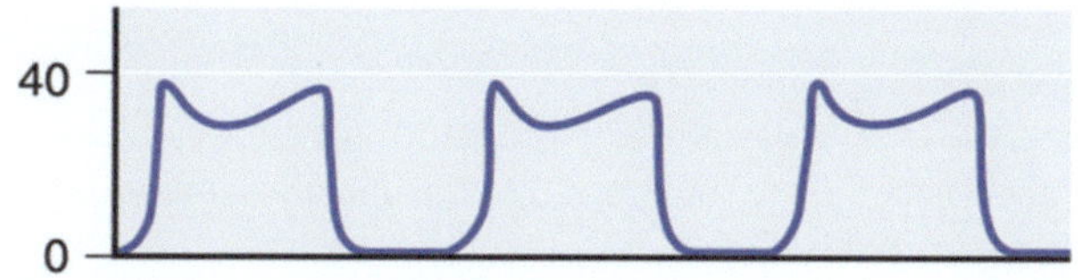

Fig. 17.1 Biphasic capnography after single lung transplant [9] (Two peaks in phase III suggest sequential emptying of two heterogeneous compartments)

mitigated by using low minute ventilation and prolonged expiratory time [8]. The resultant hypercapnia is often well tolerated in the COPD patient. Visually assess the lungs for degree of expansion and adjust ventilator settings accordingly. Differential ventilation using a separate ventilator or Ambu bag is possible but impractical for extended periods. Early extubation is ideal in these circumstances.

In the setting of restrictive lung disease, differential compliance poses less of an issue. There will be preferential ventilation of the newly transplanted lung, which should be far more compliant. Perform two lung ventilation with smaller volumes (i.e., 4cc/kg) to avoid overinflating the transplanted lung. Shunting and hypoxia may occur, and iNO can be considered to minimize FiO_2.

17.3 Lung Transplantation on Peripheral V-V ECMO

V-V ECMO as bridge to transplant is used for patients with an acute deterioration who have failed mechanical ventilation and conventional treatment. Candidacy for lung transplantation while on V-V ECMO is debated across lung transplant programs, and decided on a case-by-case basis. V-V ECMO offers only oxygenation/ventilatory support and no circulatory support. The common V-V ECMO configurations include [10]:

- **Femoro-Jugular V-V ECMO:**
  - Inflow (to oxygenator): hepatic IVC via femoral vein
  - Outflow (to patient): right atrium via right IJ
  - Benefits/issues: Minimal recirculation with proper placement
- **Femoro-Femoral V-V ECMO:**
  - Inflow: hepatic IVC via femoral vein
  - Outflow: right atrium via contralateral femoral vein
  - Benefits/issues: Prone to recirculation
- **Dual-Lumen BiCaval Cannula (e.g., Avalon):**
  - Single cannula, inflow ports in the SVC and IVC, outflow into the RA
  - Placement only in the right IJ (very stiff cannula)
  - Benefits/issues: Minimal recirculation but often positional, flows limited by narrower cannula

Induction

These patients are frequently already intubated and induction consists of starting volatile anesthetics once transferred to the anesthesia ventilator +/− ongoing IV sedation from the ICU.

Airway

Intubation with a double-lumen tube (see further below).

Lines

Most patients will have an ECMO cannula in the right IJ, given its direct course to the SVC and right atrium. Additionally, most have a triple-lumen CVC in place, commonly in the left IJ or femoral vein. As such, placement of a Cordis for PA catheter insertion will typically require:

- Placement in a vacant left IJ
- Placement in the left IJ already occupied by a CVC—a "double stick" after thoroughly prepping the site
- Placement in the left IJ by rewiring an existing CVC and exchanging it for the Cordis (after thoroughly prepping the site)
- Placement of the Cordis in the left subclavian

All these options carry risks, and the optimal strategy varies. A small caliber IJ precludes "double-sticking", and a patient dependent on several vasopressors should not have their sole CVC exchanged. Furthermore, the large ECMO cannula may preclude floating of the PA catheter. During CVC placement, close coordination with the perfusion team is necessary. ECMO flows should be reduced during insertion to avoid entraining air.

Position and Incision

Supine (arms out), clamshell thoracotomy.

Anticoagulation

IV Heparin Infusion (typical dose range 500–1000 U/hr)—titrated to PTT in ICU. ACT target 160–180 in OR.

Immunosuppression
500 mg Solumedrol at induction.

Antibiotic
Piperacillin/Tazobactam (4.5 g) q6h.

Disposition
ICU.

Operative Course

As V-V ECMO offers no cardiac support, hemodynamics during transplant will be the same as for off-ECMO transplant, requiring fluid and vasopressors to augment blood pressure. Target normotension to slight hypertension during periods of manipulation. Mediastinal manipulation may impair venous return to both the heart and ECMO inflow, compromising both hemodynamics and respiratory support. If the patient becomes hemodynamically unstable, the ECMO circuit can be reconfigured to allow for an arterial outflow cannula, creating a hybrid system [11].

Ventilate the native lungs with lung protective settings, aiming for a combined tidal volume of 200-250 cc. Although the gas exchange of the native lungs is severely impaired, the continued ventilation provides some protection against shunting or hypoxia due to inadequate ECMO flow.

V-V ECMO requires continuous anticoagulation to prevent the circuit from clotting. At TGH, ECMO patients are anticoagulated with IV heparin, with typical dose ranges of 500–1000 U/hr (10-20u/kg/hr). Obtain a baseline ACT at the start of surgery and titrate the heparin infusion accordingly, targeting an ACT of 160–180. Coordinate regular ACT checks with the perfusionist at an interval determined by ACT stability (q1-2h). Reconfiguring the ECMO circuit with an arterial cannula requires an ACT >220 prior to cannulation.

Reperfusion on peripheral V-V ECMO follows the same principles as reperfusion on V-A ECMO, apart from the ability to manipulate PA flow. The PA clamp will be removed gradually to lessen the risk of reperfusion injury. Ventilate with lung protective settings and a lower FiO_2.

An additional consideration for surgery on V-V ECMO is the disordered coagulation associated with prolonged ECMO. Prothrombotic changes include increased thrombin activation [12], contact pathway activation [13], and increased coagulation factor levels, including Factors II, V, VIII, and X [14]. Hypocoagulable changes include acquired von Willebrand syndrome [13, 15, 16], platelet dysfunction and hemolysis [13, 17, 18], increased fibrinolysis [19], consumption of coagulation factors, and ongoing systemic anticoagulation. The combined effect of these changes on coagulation is unpredictable, and conventional lab tests (CBC, coags) may be misleading. There has been growing interest in the use of thromboelastometry-based assays (i.e., ROTEM), but few studies have directly examined their utility in ECMO patients [20, 21]. The most recent Extracorporeal Life Support Organization (ELSO) Anticoagulation Guidelines [22] recommend the following transfusion targets for patients on ECMO, although they are not specifically intended to guide transfusion during lung transplant surgery (Table 17.2).

If the transplant is uneventful and the grafts are performing well, the surgical team will attempt to wean ECMO and decannulate at the end of surgery. Assessing readiness for decannulation involves gradual reductions in sweep and ECMO flow while increasing minute ventilation and FiO2. Consider decannulation if oxygenation and ventilation are adequate ($SpO_2 > 92\%$ on $FiO_2 < 50\%$) with reasonable vent settings. If oxygenation or ventilation is inadequate, the patient is kept on V-V ECMO until graft function improves.

17.4 Lung Transplantation for COVID ARDS

In late 2019, the first outbreak of Severe Acute Respiratory Syndrome Coronavirus 2 (SARS-CoV-2) occurred in Wuhan, China. Spreading quickly to Europe, and then subsequently North America, the World Health Organization (WHO) declared a global health emergency from January 2020 to May 2023.

SARS-CoV-2 produces severe respiratory symptoms, sometimes progressing to acute respiratory distress syndrome (ARDS), respiratory failure requiring hospitalization, intensive care admission, mechanical ventilation, and ECLS. This was particularly notable in the early waves when medical therapies and vaccination were unavailable. In the most severe of cases, irreversible lung damage occurs, with many patients completely reliant on mechanical ventilation and ECMO for survival. These patients also develop secondary multiorgan failure (cardiac, renal, and others) and thrombotic complications.

Lung transplantation is a viable option for patients with irreversible pulmonary damage secondary to SARS-CoV-2. The TLTP stipulated 10 considerations in the assessment of a COVID ARDS patient for lung transplantation [23]. They are summarized below:

1. Candidates should be younger than 65 years of age.
2. Candidates should only have single organ dysfunction.
3. Sufficient time should be allowed for lung recovery. Transplant should not be considered less than 4–6 weeks after initial signs of respiratory failure.
4. There should be radiologic evidence of irreversible lung disease (fibrosis, bullous disease).
5. The patient should be awake and participate in the discussion regarding transplantation.

Table 17.2 ELSO anticoagulation guideline transfusion targets

	Goal	Product to transfuse
Platelets	≥100,000 × 10^9/L (bleeding patient) ≥50,000–100,000 × 10^9/L (nonbleeding patient)	Platelets 10 ml/kg (max 2 units)
INR	<1.5 (bleeding patient) <3 (nonbleeding patient)	Fresh frozen plasma 10 ml/kg (max 2 units)
Fibrinogen	>1.5 g/L (bleeding patient or before surgical intervention) >1 g/L (nonbleeding patient)	Cryoprecipitate 1 unit/5 kg (max 6 units)
Hemoglobin	>70–90 g/L (consider higher goal for neonates and children with cyanotic congenital heart disease or lower goal for stable, adult patients)	Packed RBCs 10 ml/kg (max 2 units)
Antithrombin	>50%-80% (>0.5–0.8 U/ml), consider AT replacement if on maximum dose of UFH and unable to obtain anticoagulation goals	AT concentrate: $\text{Thrombate III dose}(\text{IU}) = \frac{(\text{desired AT} - \text{current AT}) \times \text{weight}(\text{kg})}{1.4}$

AT antithrombin, *RBCs* red blood cells, *INR* international normalized ratio, *UFH* unfractionated heparin

6. Patients should be able to participate in physical rehabilitation while awaiting transplantation.
7. Candidates should adhere to the usual criteria for transplantation (i.e., BMI, absence of other comorbid illness such as CAD).
8. Candidates should have a recent negative SARS-CoV-2 PCR result.
9. Transplantation should be done at a center with experience in high-risk transplantation.
10. The transplant center should have a broad donor pool and low waiting list mortality—this ensures equitable organ allocation and ensures that organs are provided to those who have the greatest likelihood of survival.

The first COVID ARDS transplant done in Canada was performed by the TLTP in February 2021. Overall, in 2021, 5 transplants for COVID ARDS were performed by the TLTP. Additionally, a case series involving 3 of the recipients was published [24], showing good short-term outcomes in these patients who were otherwise ECMO-dependent. Overall, outcomes in ECMO-dependent COVID ARDS patients undergoing transplantation have been positive [25], due in large part to careful candidate selection.

The early transplant experience with COVID ARDS involved those with severe infection and irrecoverable lung function requiring long-term ECMO and/or mechanical ventilation. More recently, as the pandemic has progressed, pulmonary fibrosis has been identified in many COVID-19 survivors. One meta-analysis of 2018 COVID survivors revealed that approximately 44.4% develop pulmonary fibrosis [26]. Fibrosis tends to be more prevalent in those with more severe infection. The time course of fibrosis is variable, with some patients showing evidence while in hospital, while others deteriorate weeks after discharge. The majority of patients show some resolution over time, however in some patients, the fibrosis is persistent [27]. A number of these patients develop functional decline and persistent oxygen requirements and are considered for listing.

Thrombotic complications were reported in COVID patients across several countries, with arterial and venous thromboses more common with severe COVID infections requiring hospitalization [28]. Patients supported on ECMO were subject to both an increased risk for both thrombotic and bleeding complications [29–31]. As such, anticoagulation management remains challenging in these patients. There are no consensus guidelines for the management of anticoagulation in the patient with COVID supported on ECMO, in general or for lung transplantation. Our institutional guidelines for management of anticoagulation on ECMO remain the same for those with severe COVID ARDS, and adjustments for thrombotic or bleeding concerns made on a case-by-case basis.

Point-of-care coagulation testing such as ROTEM has been useful in demonstrating the hypercoagulable state of some COVID ARDS patients. In this group, the amplitude at 10 minutes (A10) and maximum clot firmness (MCF) are greatly exaggerated. The FIBTEM also shows hypofibrinolysis [32, 33]. In COVID-ARDS patients for lung transplantation supported on ECMO, it is advisable to obtain a baseline ROTEM to assess for both coagulopathy and prothrombotic states warranting strict ACT monitoring.

The anesthetic management of these patients will depend upon the severity and state of the patient. In those with severe COVID ARDS reliant on ECMO, the management will follow the principles outlined in Sect. 17.3. In those who recover to be weaned off ECMO, their clinical management and considerations are similar to those with other fibrotic lung disease. Because of the fibrotic and non-suppurative nature of COVID ARDS, candidates are considered for both double and single lung transplantation.

17.5 Combined Cardiac Surgery and Lung Transplantation

Concomitant cardiac surgery at the time of lung transplantation is an uncommon occurrence at the TLTP, with fewer than 10 cases per year. Indications for combined cardiac surgery at the time of lung transplantation include [34]:

1. Coronary artery disease
2. Valvular heart disease (left and right-sided)
3. Patent foramen ovale or atrial septal defect with significant shunt
4. Arrhythmia surgery (maze procedure, ligation of atrial appendage)
5. Pulmonary artery dilatation requiring PA reconstruction

The decision to allow for concomitant cardiac surgery depends on program experience, as there is increased potential for morbidity and mortality associated with combined procedures. These concerns arise from the need for full cardiopulmonary bypass and heparinization, the extended procedure time, and the raised potential for cardiac dysfunction in this patient population.

The TLTP typically allows for concomitant coronary artery bypass grafting and lung transplantation in patients where percutaneous coronary intervention is not possible (or may delay necessary transplantation due to dual antiplatelet requirements). This is particularly favored when the coronary anatomy is straightforward and cardiac function is preserved, as the literature shows positive outcomes [35–37].

Additional Equipment
Consideration for TEE where appropriate (valvular or structural interventions, unclear cardiac function).

Pre-induction
Apply defibrillator pads and ensure capture prior to induction, particularly in patients with left main and/or multivessel disease. The surgeons will likely remove these during preparation for the clamshell incision.

Induction
Both cardiac and thoracic surgery staff should be present at the time of induction in the event hemodynamic instability arises. This allows for expedient chest entry and cannulation for the appropriate mode of ECLS. Induction for the patient with significant coronary artery disease should avoid prolonged hypotension or tachycardia, which can aggravate myocardial ischemia.

Airway
Intubation with a double-lumen tube.

Lines
Conventional line placement.

Position and Incision
Supine (arms out), clamshell thoracotomy.

Anticoagulation
Full CPB dosing (400–600IU of heparin per Kg to achieve an ACT >480 s).

Immunosuppression
500 mg Solumedrol at induction.

Antibiotic
Piperacillin/Tazobactam (4.5 g) q6h.

Disposition
ICU.

Operative Course [Specifically for CABG]

Following induction, the lung transplant surgeons will perform a full clamshell incision to expose the heart and lungs, as in a conventional double lung transplant. Typically, the cardiac surgery procedure will take place first, followed by the lung transplant.

The mode of ECLS is dependent on the cardiac procedure and surgical preference. Full CPB is required for complex coronary grafting (multiple vessels, less accessible disease). Once the chest and pericardium are open, the patient is fully heparinized (ACT >480s) for aortic and venous cannulation, and bypass is instituted. Cross-clamping of the aorta and delivery of cold cardioplegia arrests the heart. Ventilation of the lungs may be stopped or continued at the preference of the surgeons.

For CABG, the use of a clamshell incision typically renders the internal mammary arteries less suitable for bypass grafting (inadequate length), thus saphenous vein conduits are generally used.

Once the cardiac intervention has been completed, the cross-clamp is removed, and the heart is reperfused. Defibrillation and/or pacing may be required to restore a perfusing rhythm. At this point, the lung transplant procedure may begin. Depending on the extent of cardiac surgery, as well as surgical preference, a decision will be made as to whether to perform the lung transplant on CPB, or to convert to ECMO.

Continuing CPB avoids circuit exchange and its associated risks. However, CPB also requires full systemic anticoagulation and significant fluid administration, which increases risk for bleeding, transfusion, and lung injury.

Conversion to ECMO affords hemodynamic support with lower anticoagulation requirements and reduced bleeding risk. The process of converting to ECMO involves weaning from cardiopulmonary bypass. Once off CPB, the aortic and venous cannulae are clamped and the CPB circuit is disconnected. A pre-primed ECMO circuit can then be attached, obviating the need for denovo cannulation.

The process of converting to ECMO carries some additional considerations, including dilution from the crystalloid-primed ECMO circuit, as well as loss of the residual blood volume that remains in the CPB circuit. The latter can be reclaimed via the cell salvage, but this is not immediate. As such, this entire process may produce periods of hypovolemia, anemia, and dilutional coagulopathy (recall that the cell saver returns red cells, but not clotting factors). Blood product transfusion may be required at this point.

Once on ECMO, the ACT can be allowed to drift down to the typical ECMO range (160–180s). In the setting of significant coagulopathy, point-of-care coagulation testing should be performed, and consideration for low-dose protamine administration can be considered if the ACT remains in the CPB range.

In patients with limited and accessible coronary disease (such as single-vessel LAD disease), the CABG may be performed on ECMO. This requires a cardiac surgeon experienced with "beating heart" bypass surgery.

17.6 Heart Lung Transplantation

These are among the rarest cases performed at TGH, as we average 1–2 per year. In the TLPT, the lung transplant surgeon is the primary surgeon, but this is a joint operation performed with the cardiac surgeons.

Indications
Congenital cardiac disease, fixed pulmonary hypertension second to cardiac disease.

Additional Equipment
TEE, Continuous Oximetry Swan, HemoSphere CO monitor. Strongly consider a brachial arterial line for these cases, which better reflect central aortic pressures and are less prone to artifactual reading, obstruction or dampening upon discontinuation of CPB.

Pre-induction
Awake central access should be considered for most of these cases. Apply defibrillator pads and ensure capture prior to induction, although the surgeons will likely remove these during preparation for the clamshell incision.

Induction
The surgeons should be in the room for induction, and a clear plan should be in place in the event of severe hemodynamic instability. In general, you can adapt the same hemodynamically conscious induction strategies taken for patients with severe pulmonary hypertension. The only caveat being that some cardiac lesions may not tolerate elevated heart rates (ischemic cardiomyopathy, end-stage HOCM), so determine whether sympathomimetics and inotropes should be avoided prior to beginning your induction. Likewise, your rescue medication must be carefully selected.

Airway
Single lumen ETT for entirety of case.

Position
Supine, arms out.

Incision
Clamshell.

Anticoagulation
Full CPB dosing (400–600 IU of heparin per Kg to achieve an ACT >480s).

Immunosuppression
1 g Solumedrol at induction (given by anesthesiologist), 500 mg Solumedrol after release of aortic cross clamp (given by perfusionist).

Antibiotic
Piperacillin/Tazobactam (4.5 g) q6h.

CPB Details
Contrary to conventional heart surgery, patient is kept warm (36–37 degrees) throughout the bypass run.

Disposition
ICU.

Operative Course

Following induction, the thoracic surgeons will perform the clamshell incision, dissection, and exposure, as in conventional double lung transplant. The patient may have high inotrope and vasopressor requirements prior to CPB.

The cardiac surgeons will cannulate the aorta, SVC and IVC. Target an SBP of 80-100 mmHg during aortic cannulation to minimize the risk of dissection. The perfusionist will ensure the ACT is over 480s before commencing bypass.

The bypass run is typically uneventful, but the perfusionists may require assistance with the blood pressure in vasoplegic patients.

Remember to pull back your PA catheter to 20 cm or less, as the catheter may be damaged or sutured by the surgeons while performing the SVC or PA anastomosis.

The surgical order of operation is tracheal anastomosis, aortic anastomosis, SVC and IVC anastomoses, and PA anastomosis.

Bronchoscopy should be performed upon completion of the tracheal anastomosis to inspect the suture line and perform bronchopulmonary toilet of the lungs.

The surgeons will release the aortic cross clamp as soon as the aortic anastomosis is complete, which ends ischemic time and begins reperfusion time. Ideally, total reperfusion time would be at least ¼ the ischemic time, or roughly 1 hour in most cases. However, this is rarely ever achieved, and reperfusion times of 20–30 minutes are more typical.

The surgeons will instruct you when to recruit and ventilate the lungs on typical lung transplant settings (i.e., 10cmH$_2$O/10 breaths per minute/ PEEP 5cmH2O, FIO2 30%). Discontinuation from bypass should occur at an FiO2 of 80%, with down-titration of FiO2 as soon as possible, using ABGs to titrate to a PaO2 of 80–100.

iNO benefits both the lungs and right ventricle and may be considered even if the lungs are performing well from an oxygenation perspective.

The choice and dose of vasoactive medications will vary depending on cardiac etiology, ischemic time, and staff preference. Remember that a short lung ischemic time (6 hrs), represents an incredibly long cardiac ischemic time (preferably under 4 hours).

Upon discontinuation of bypass, TEE will be performed to assess biventricular function as well as the SVC, IVC, and aortic anastomoses. Pulmonary vein flow is less of a concern than in lung transplantation as all the attachments are native. A stenotic SVC with turbulent flow may predispose to SVC syndrome, so assess for facial plethora and airway edema if the suspicion is high.

Coagulopathy is more of an issue than in typical lung transplants. Be prepared to check coags/ROTEM and transfuse as appropriate.

17.7 Redo Lung Transplant

Redo lung transplantation remains the only treatment option for lung transplant recipients who develop significant chronic lung allograft dysfunction (CLAD). Given the limited availability of grafts, redo lung transplantation is uncommon, and the TLTP performs fewer than 10 per year.

There have been two described phenotypes of CLAD:

- Bronchiolitis Obliterans Syndrome (BOS): Obstructive pulmonary function, air trapping on imaging and obliterative bronchiolitis on pathology.
- Restrictive CLAD (rCLAD), also referred to as Restrictive Allograft Syndrome (RAS): Restrictive pulmonary function, persistent pleuro-parenchymal infiltrates on imaging, and pleuroparenchymal fibroelastosis on pathology.
- There can be significant heterogeneity and overlap.

Compared to BOS, rCLAD has a worse survival rate following diagnosis (6–18 months vs 3–5 years) [38] and worse survival after retransplantation [39]. In the setting of CLAD, both single and double lung transplantation have been offered.

As compared with primary lung transplantation, there is heterogeneity among the literature, with some studies reporting worse survival (as compared to primary lung transplantation) in CLAD patients [40, 41], whereas others report comparable survival at 1, 2, and 5 years post-retransplant [42–44] (particularly in BOS patients).

Other less common indications for redo lung transplantation include severe airway complications post-transplant, and primary graft dysfunction [45]. Retransplantation in the setting of severe PGD is associated with poorer survival [43].

Additionally, management and planning for patients presenting for redo transplantation must also consider the indication for the primary lung transplantation, which may incorporate extrapulmonary disease involvement that may still be relevant or have further progressed.

Additional Equipment
None specific.

Induction
Dependent on clinical status and stability, however, commonly follow routine induction as for primary double lung transplantation.

Airway
Intubation with a double-lumen tube. In the rare circumstance of retransplantation for airway-related concerns, special consideration and planning may be required for anatomical abnormalities (e.g., stenosed/narrowed airways).

Lines
Conventional line placement.

Position and Incision
Supine (arms out), clamshell thoracotomy.

Immunosuppression
500 mg Solumedrol at induction.

Antibiotic
Piperacillin/Tazobactam (4.5 g) q6h.

Disposition
ICU.

Operative Course
While the sequence and conduct of redo transplantation follows the same principles as primary double (or single) lung transplant procedures, there are several added challenges. Given the underlying lung dysfunction, oxygenation and CO_2 may be severely compromised, and ventilation may be challenging due to the presence of obstructive or restrictive lung physiology.

From a surgical perspective, adherence of the transplanted lungs to the chest wall increases the risk for lung injury during dissection, increased bleeding, need for more aggressive retraction, and overall increase in surgical time. Because of these challenges, a second staff surgeon is often present for these procedures.

The use of V-A ECMO may lessen many of these challenges, and is often (but not always) used for redo lung transplantation. The decision to use ECMO will weigh on the ability to safely manage oxygenation and hemodynamics. However, due to the increased amount of surgical dissection, the use of ECMO can exacerbate bleeding and need for transfusion. Additionally, institution of ECMO itself may be more challenging for the same reasons. Redo transplantation is associated with increased transfusion, both locally and in the lung transplant literature.

References

1. Yeung JC, et al. Lung transplantation for cystic fibrosis. J Heart Lung Transplant. 2020;39(6):553–60.
2. Hauser N, Orsini J. Cepacia syndrome in a non-cystic fibrosis patient. Case Rep Infect Dis. 2015;2015:537627.
3. Shah P, et al. Cystic fibrosis foundation consensus statements for the care of cystic fibrosis lung transplant recipients. J Heart Lung Transplant. 2021;40(7):539–56.
4. Kang SH, et al. Chronic rhinosinusitis and nasal polyposis in cystic fibrosis: update on diagnosis and treatment. J Bras Pneumol. 2015;41(1):65–76.
5. Ranganath NK, et al. Single and double lung transplantation have equivalent survival for idiopathic pulmonary fibrosis. Ann Thorac Surg. 2020;109(1):211–7.
6. Crawford TC, et al. Five-year mortality Hazard is reduced in chronic obstructive pulmonary disease patients receiving double- versus single-lung transplants. J Surg Res. 2019;237:118–25.
7. Siddiqui FM, Diamond JM. Lung transplantation for chronic obstructive pulmonary disease: past, present, and future directions. Curr Opin Pulm Med. 2018;24(2):199–204.
8. Martin AK, et al. Anesthetic management of lung transplantation: impact of presenting disease. Curr Opin Anaesthesiol. 2020;33(1):43–9.

9. Chitilian H, et al. Respiratory monitoring. In: Miller R, editor. Miller's anesthesia. Elsevier; 2015.
10. Banfi C, et al. Veno-venous extracorporeal membrane oxygenation: cannulation techniques. J Thorac Dis. 2016;8(12):3762–73.
11. Brasseur A, et al. Hybrid extracorporeal membrane oxygenation. J Thorac Dis. 2018;10(Suppl 5):S707–S15.
12. Cartwright B, et al. Hemostasis, coagulation and thrombin in venoarterial and venovenous extracorporeal membrane oxygenation: the HECTIC study. Sci Rep. 2021;11(1):7975.
13. Doyle AJ, Hunt BJ. Current understanding of how extracorporeal membrane oxygenators activate Haemostasis and other blood components. Front Med (Lausanne). 2018;5:352.
14. Mazzeffi M, et al. Coagulation factor levels and underlying thrombin generation patterns in adult extracorporeal membrane oxygenation patients. Anesth Analg. 2019;129(3):659–66.
15. Kalbhenn J, et al. Acquired von Willebrand syndrome and impaired platelet function during venovenous extracorporeal membrane oxygenation: rapid onset and fast recovery. J Heart Lung Transplant. 2018;37(8):985–91.
16. Panholzer B, et al. Acquired von Willebrand syndrome in ECMO patients: a 3-year cohort study. Blood Cells Mol Dis. 2021;87:102526.
17. Jiritano F, et al. Platelets and extra-corporeal membrane oxygenation in adult patients: a systematic review and meta-analysis. Intensive Care Med. 2020;46(6):1154–69.
18. Panigada M, et al. Hemostasis changes during veno-venous extracorporeal membrane oxygenation for respiratory support in adults. Minerva Anestesiol. 2016;82(2):170–9.
19. Malfertheiner MV, et al. Hemostatic changes during extracorporeal membrane oxygenation: a prospective randomized clinical trial comparing three different extracorporeal membrane oxygenation systems. Crit Care Med. 2016;44(4):747–54.
20. Nair P, et al. Prospective observational study of hemostatic alterations during adult extracorporeal membrane oxygenation (ECMO) using point-of-care thromboelastometry and platelet aggregometry. J Cardiothorac Vasc Anesth. 2015;29(2):288–96.
21. Panigada M, et al. Thromboelastography-based anticoagulation management during extracorporeal membrane oxygenation: a safety and feasibility pilot study. Ann Intensive Care. 2018;8(1):7.
22. McMichael ABV, et al. 2021 ELSO adult and pediatric anticoagulation guidelines. ASAIO J. 2022;68(3):303–10.
23. Cypel M, Keshavjee S. When to consider lung transplantation for COVID-19. Lancet Respir Med. 2020;8(10):944–6.
24. Yeung JC, et al. Lung transplantation for acute COVID-19: the Toronto lung transplant program experience. CMAJ. 2021;193(38):E1494–E97.
25. Bharat A, et al. Early outcomes after lung transplantation for severe COVID-19: a series of the first consecutive cases from four countries. Lancet Respir Med. 2021;9(5):487–97.
26. Hama Amin BJ, et al. Post COVID-19 pulmonary fibrosis; a meta-analysis study. Ann Med Surg (Lond). 2022;77:103590.
27. Liu M, et al. Follow-up study of the chest CT characteristics of COVID-19 survivors seven months after recovery. Front Med (Lausanne). 2021;8:636298.
28. Burn E, et al. Venous or arterial thrombosis and deaths among COVID-19 cases: a European network cohort study. Lancet Infect Dis. 2022;22(8):1142–52.
29. Yusuff H, et al. Thrombosis and coagulopathy in COVID-19 patients rceiving ECMO: a narrative review of current literature. J Cardiothorac Vasc Anesth. 2022;36(8 Pt B):3312-17
30. Ripoll B, et al. Observational study of thrombosis and bleeding in COVID-19 VV ECMO patients. Int J Artif Organs. 2022;45(2):239–42.
31. Schmidt M, et al. Extracorporeal membrane oxygenation for severe acute respiratory distress syndrome associated with COVID-19: a retrospective cohort study. Lancet Respir Med. 2020;8(11):1121–31.
32. Boss K, et al. Pathological findings in rotation thromboelastometry associated with thromboembolic events in COVID-19 patients. Thromb J. 2021;19(1):10.

33. Mitrovic M, et al. Rotational thromboelastometry (ROTEM) profiling of COVID-19 patients. Platelets. 2021;32(5):690–6.
34. Biniwale R, et al. Lung transplantation and concomitant cardiac surgery: is it justified? J Thorac Cardiovasc Surg. 2016;151(2):560–6.
35. Patel VS, et al. Clinical outcome after coronary artery revascularization and lung transplantation. Ann Thorac Surg 2003;75(2):372–7; discussion 77.
36. Lee R, et al. Concomitant coronary artery revascularization to allow successful lung transplantation in selected patients with coronary artery disease. J Thorac Cardiovasc Surg. 2002;124(6):1250–1.
37. Sherman W, et al. Lung transplantation and coronary artery disease. Ann Thorac Surg. 2011;92(1):303–8.
38. Verleden SE, et al. Chronic lung allograft dysfunction phenotypes and treatment. J Thorac Dis. 2017;9(8):2650–9.
39. Verleden SE, et al. Impact of CLAD phenotype on survival after lung Retransplantation: a multicenter study. Am J Transplant. 2015;15(8):2223–30.
40. Ren D, et al. Retransplantation outcomes at a large lung transplantation program. Transplant Direct. 2018;4(11):e404.
41. Osaki S, et al. Redo lung transplantation for acute and chronic lung allograft failure: long-term follow-up in a single center. Eur J Cardiothorac Surg. 2008;34(6):1191–7.
42. Wallinder A, et al. Outcomes and long-term survival after pulmonary Retransplantation: a single-center experience. Ann Thorac Surg. 2019;108(4):1037–44.
43. Aigner C, et al. Pulmonary retransplantation: is it worth the effort? A long-term analysis of 46 cases. J Heart Lung Transplant. 2008;27(1):60–5.
44. Strueber M, et al. Long-term outcome after pulmonary retransplantation. J Thorac Cardiovasc Surg. 2006;132(2):407–12.
45. Novick RJ, et al. Redo lung transplantation: a north American-European experience. J Heart Lung Transplant. 1993;12(1 Pt 1):5–15; discussion 15–6.

Index

M. Salvatori, A. Huang (eds.), *Anesthesia for Lung Transplant*,
https://doi.org/10.1007/978-3-032-06112-6

MIX
Papier aus verantwortungsvollen Quellen
Paper from responsible sources
FSC® C105338

If you have any concerns about our products,
you can contact us on
ProductSafety@springernature.com

In case Publisher is established outside the EU,
the EU authorized representative is:
Springer Nature Customer Service Center GmbH
Europaplatz 3, 69115 Heidelberg, Germany

Printed by Libri Plureos GmbH
in Hamburg, Germany